CLASSIC

MATHEMAGIC

CLASSIC

MATHEMAGIC

Raymond Blum, Adam Hart-Davis,
Bob Longe, & Derrick Niederman

MetroBooks

©2002 by Sterling Publishing Co., Inc.
Published by MetroBooks by arrangement with Sterling Publishing Co., Inc.
First MetroBooks edition 2002

Library of Congress Cataloging-in-Publication Data Available Upon Request
ISBN 1-58663-683-9

10 9 8 7 6 5 4 3 2 1

Material in this book previously appeared in Mathamusements, ©1997 by
Raymond Blum; The Magical Math Book, ©1997 by Bob Longe; Amazing Math
Puzzles, ©1998 by Adam Hart-Davis; The Little Giant Book of Math Puzzles, ©2000
by Derrick Niederman; Hard-to-Solve Math Puzzles, ©2001 by Derrick Niederman;
all published by Sterling Publishing Co., Inc.

For bulk purchases and special sales, please contact:
MetroBooks
Attention: Sales Department
230 Fifth Avenue
New York, NY 10001
212/685-6610†FAX 212/685-3916

Visit our website: www.metrobooks.com

CLASSIC MATHEMAGIC

Contents

Introduction vii

Glossary 1

Tricks of the Trade 9

Child's Play? 19

Working Towards Wizardry 83

Magical Math 117

Great Math Challenges 181

Answers 215

Index 277

INTRODUCTION

Numbers surround us.

What page? How old? What date? How much/many? How fast? When?

This book is packed with mathematical puzzles of every type—amazing number tricks, beautiful geometric designs, challenging puzzles, marvelous memory tricks, and many other mathematical amusements. Some of the puzzles are fairly easy, but others are quite challenging. Just remember, the difficulty level isn't the same for every solver out there, and each mind excels at its own specialty. The variety in this book will expand your mathematical awareness and understanding, while at the same time being lots of fun.

The math principles used in this book are explained early on, in various simple tricks and examples, and the book is organized with a glossary to familiarize readers with terms used in the book. As you proceed, you'll see how the principles are applied in different ways. This way, you are provided with something of a sequential approach to these principles.

Some puzzles in this book have hints, whether you need them or not. You'll find the hints printed upside-down below those puzzles. Remember, using a hint is perfectly okay if a particular puzzle stumps you. Sometimes the hints help you understand precisely what the question is asking, and other times they lead you on the way to the answer.

One last thought. If some of these puzzles elude you, don't get discouraged! You're not expected to get every one of them the first time around. Many are devised to introduce you to brand new ways of thinking or approaches that you may not have seen before. We have tried to give you enough information in the Answers section so that the next time you come across a similar kind of puzzle, you'll think it's a piece of cake. By the time you've finished this book—or even before—you will be thinking like a real puzzle solver.

Are you ready to get started? Sharpen your pencils, boot up your brain, and have fun!

GLOSSARY

algebra A mathematical language that uses letters along with numbers. $5x + 6 = 21$ is an example of an algebra problem.

approximately equal to (») A symbol used when an answer is close to an exact answer.

area The amount of space inside a figure.

average The sum of a set of numbers divided by how many numbers there are.

billion A word name for 1,000,000,000.

binary number system A number system based on the number 2.

binomial An algebraic expression that has two terms. Example: $2x + 1$

birth date The date of the day that you were born.

Celsius The temperature scale of the metric system.

center of a circle The point that is the same distance from all of the points on a circle.

centimeter A metric unit of length that approximately equals .4 of an inch.

circumference The distance around a circle.

circumference of the Earth Approximately 24,902 miles (40,075 km).

compass An instrument used to draw circles.

compass points North, East, South, and West.

cube A three-dimensional figure with six square faces all the same size.

day A unit of time equal to 24 hours.

decimal part of a number The digits to the right of the decimal point.

diameter The distance across the center of a circle.

diameter of the Earth Approximately 7,927 miles (12,757 km).

digit Any of the symbols 0 to 9 used to write numbers. Example: 6,593 is a four-digit number.

distance formula distance = rate $\times$ time

estimate To give an approximate rather than an exact answer.

even numbers The numbers 0, 2, 4, 6, 8, 10....

Fahrenheit The temperature scale of the U.S. system.

formula An algebraic sentence that states a math fact or rule. Example: The area of a rectangle equals the length times the width. ($A = l \, w$).

geometric Consisting of straight lines, circles, angles, triangles, etc.

geometry A kind of mathematics that studies points, lines, angles, and different shapes.

gravity The force that pulls things downward.

gravity factor The number that you multiply your Earth weight by to find your approximate weight at different places in our solar system.

grid Horizontal and vertical parallel lines in a checkerboard pattern.

hexagon A polygon with six sides.

horizontal line A line that runs straight across from left to right.

hour A unit of time equal to 60 minutes.

hundreds place Example: In the number 8,376, the 3 is in the hundreds place.

inch A U.S. unit of length equal to 2.54 centimeters.

is greater than (>) A symbol used to compare two numbers when the larger number is written first. Example: 73 > 5

is less than (<) A symbol used to compare two numbers when the smaller number is written first. Example: 12 < 47

kilogram A metric unit of mass that approximately equals 2.2 pounds.

kilometer A metric unit of length that approximately equals .6 of a mile.

leap year A year having 366 days. A leap year is a year that can be divided by 4 exactly. Examples: 1996, 1992, 1988, 1984, 1980, etc.

light (speed of) Approximately 186,000 miles per second (300,000 km/sec).

line design A geometric design made with straight lines.

mathemagic Magic tricks that use numbers.

meteorologist A person who studies and reports the weather.

mile A U.S. unit of length that equals approximately 1.6 kilometers.

million A word name for 1,000,000.

minute A unit of time equal to 60 seconds.

mirror symmetry When one half of a figure is the mirror image of the other half.

mnemonic A word, phrase, rhyme, or anything that can be used to help you remember.

numerology Assigns everyone a number based on his or her name or birth date. This number might reveal information about personality.

odd numbers The numbers 1, 3, 5, 7, 9, 11....

ones place Example: In the number 8,376, the 6 is in the ones place.

operations $+, -, \times,$ and $\div$

order of operations Rules about the order in which operations should be done. 1. Parentheses 2. Exponents 3. From left to right, multiplications and divisions 4. From left to right, additions and subtractions.

origami A Japanese word that means "the folding of paper."

palindrome Any group of letters or numbers that reads the same forward and backward.

parallel lines Lines in the same plane that never intersect.

parallelogram A quadrilateral with two pairs of parallel sides.

perimeter The distance around the rim or border of a polygon.

pi (π) The number obtained by dividing the circumference of a circle by its diameter. It approximately equals 3.14 .

polygon A closed two-dimensional figure with three or more sides.

powers of 2 Each number is multiplied by 2 to get the next number. 1, 2, 4, 8, 16....

pyramid A three-dimensional figure whose base is a polygon and whose faces are triangles with a common vertex.

quadrilateral A polygon with four sides.

quadrillion A word name for a 1 with 15 zeroes after it: 1,000,000,000,000,000.

quintillion A word name for a 1 with 18 zeroes after it: 1,000,000,000,000,000,000.

radius The distance from the center of the circle to any point on the circle.

rate The speed of an object.

rectangle A parallelogram with four right angles.

remainder The number left over after dividing.

repeating decimal A decimal in which a digit or group of digits to the right of the decimal point repeats forever. Example: 17.333333333....

right angle An angle that has a measure of 90 degrees.

sequence A set of numbers in a certain pattern or order. Example: 3, 6, 9, 12....

setup When cards or props are arranged before performing a magic trick.

similar figures Figures that have the same shape but may not have the same size.

sound (speed of) Approximately 1,100 feet per second (330 m/sec).

square A parallelogram with four right angles and four equal sides.

squaring a number When a number is multiplied by itself. Example: $7 \times 7 = 49$

sum The answer to an addition problem.

symmetry What a shape has when it can be folded in half and the two halves match exactly.

tablespoon A U.S. unit of measure equal to three teaspoons.

tangram puzzle A seven-piece puzzle that can be put together to make hundreds of different shapes and figures.

tans The seven puzzle pieces of a tangram puzzle.

teaspoon A U.S. unit of measure equal to ⅓ of a tablespoon.

tens place Example: In the number 8,376, the 7 is in the tens place.

thousand A word name for 1,000.

thousands place Example: In the number 8,376, the 8 is in the thousands place.

topology A kind of mathematics that studies shapes and what happens to those shapes when they are folded, pulled, bent, or stretched out of shape.

triangle A polygon with three sides.

trillion A word name for 1,000,000,000,000.

vertex (plural: vertices) The point where lines meet to form an angle.

vertical line A line that runs straight up and down.

year A unit of time equal to 365 or 365.25 days.

TRICKS OF
THE TRADE

MNEMONICS

Is Less Than (<), Is Greater Than (>)

The symbols are formed by your two hands. Most people use their left hand *less* and their right hand *more* (greater).

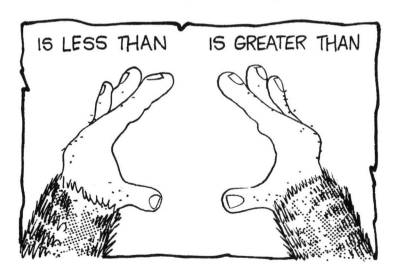

IS LESS THAN   IS GREATER THAN

How Many Teaspoons in a Tablespoon?

Both teaspoon and tablespoon start with the letter t. What number rhymes with t?...3!

3 teaspoons = 1 tablespoon

The Order of Operations in Long Division

			Example	**79 ÷ 3**
<u>D</u>ad	<u>D</u>ivide	<u>D</u>ivide 7 by 3.		26
<u>M</u>om	<u>M</u>ultiply	<u>M</u>ultiply 2 by 3.		3)79
<u>S</u>ister	<u>S</u>ubtract	<u>S</u>ubtract 6 from 7.		6
<u>B</u>rother	<u>B</u>ring down	<u>B</u>ring down the 9.		19
<u>R</u>over	<u>R</u>estart or	<u>R</u>estart — Divide 19		18
	<u>R</u>emainder	by 3, and so on.		1

The Clockwise Order of Compass Points

<u>N</u>ever <u>E</u>at <u>S</u>hredded <u>W</u>heat, or
<u>N</u>ever <u>E</u>at <u>S</u>oggy <u>W</u>affles, or
<u>N</u>ever <u>E</u>at <u>S</u>limy <u>W</u>orms, or
<u>N</u>ever <u>E</u>at <u>S</u>eptic <u>W</u>aste

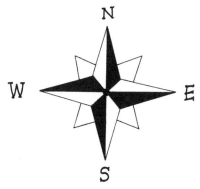

Dividing Fractions

"If it's a fraction you are dividing by, turn it upside down and multiply."

Example

$$\frac{3}{8} \div \frac{1}{2} \qquad \frac{1}{2} \text{ turned upside down is } \frac{2}{1}$$

$$\text{So } \frac{3}{8} \div \frac{1}{2} = \frac{3}{8} \times \frac{2}{1} = \frac{6}{8} = \frac{3}{4}$$

Perimeter

Perimeter is the distance around the <u>rim</u> (border, edge, boundary) of a surface or figure.

The Distance Formula

d i r t

⟶ distance is equal to rate × time

Pi (π) Rounded to Ten Decimal Places (3.1415926536)

The number of letters in each word reveals each digit.

May I have a large container of orange juice
3 .1 4   1   5   9   2   6   5
now please ?
3   6

The Order of Operations

		Example
Please	Parentheses	$20 \div 5 + \underline{(6 - 4)} \times 3^2$
Excuse	Exponents	$20 \div 5 + 2 \times \underline{3^2}$
My Dear	Multiplications and Divisions	$\underline{20 \div 5} + \underline{2 \times 9}$
Aunt Sally	Additions and Subtractions	$\underline{4 + 18}$
		22

Trigonometric Ratios

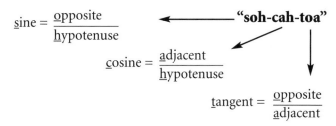

$$\text{\underline{s}ine} = \frac{\text{\underline{o}pposite}}{\text{\underline{h}ypotenuse}}$$

"soh-cah-toa"

$$\text{\underline{c}osine} = \frac{\text{\underline{a}djacent}}{\text{\underline{h}ypotenuse}}$$

$$\text{\underline{t}angent} = \frac{\text{\underline{o}pposite}}{\text{\underline{a}djacent}}$$

Multiplying Two Binomials

FOIL   Multiply the <u>F</u>irst terms, the <u>O</u>utside terms, the <u>I</u>nside terms, and the <u>L</u>ast terms.

Example

$$\begin{array}{cccc} \text{\underline{F}} & \text{\underline{O}} & \text{\underline{I}} & \text{\underline{L}} \end{array}$$
$$(x + 3)\,(x + 4) \;=\; x{\cdot}x + 4{\cdot}x + 3{\cdot}x + 3{\cdot}4 \;=$$
$$x2 + 4x + 3x + 12 \;=\; x2 + 7x + 12$$

Very Fast Multiplying

You can amaze your family and friends by multiplying large numbers in your head. It is easy to do when you know the secret shortcuts!

Multiplying a Two-Digit Number by 11

Example 26×11

What to Do

1. Separate the two digits.   **2 _ 6**
2. Add the two digits together.   **2 + 6 = 8**
3. Put that sum between the two digits.   **2<u>8</u>6**

So $26 \times 11 = 286$

If the sum of the two digits is greater than 9, you have to carry a 1 and add it to the first digit.

Example 84 × 11

$$(8 + 4 = 12) \qquad\qquad 8\ \underline{12}^*\ 4$$

* The sum of the digits is greater than 9, so carry the 1 and add it
  to the 8. The final answer is **924**.

The Secret

Multiply 26 × 11 to see why this trick works.

```
              2 6
            × 1 1
              2 6
            2 6
first digit ──► 2 8 6 ◄── second digit
              ▲
          2 + 6 = 8
```

Squaring a Two-Digit Number Ending
in the Number 5

("Squaring" a number means multiplying a number by itself.)

Example   75 × 75

What to Do

1. Multiply the first digit by one more
   than itself. (One more than **7** is **8**).   **7 × 8 = 56**

2. Put **25** after that answer (the
   answer will always end in 25).   **5625**

$$\textbf{So 75} \times \textbf{75} = \textbf{5625}$$

The Secret

This trick uses an algebraic procedure called squaring a binomial, and only works when the number you are squaring ends in 5.

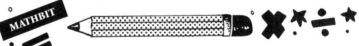

Add numbers to make squares:

1=	$1 = 1 \times 1$
1+**2**+1=	$4 = 2 \times 2$
1+2+**3**+2+1=	$9 = 3 \times 3$
1+2+3+**4**+3+2+1=	$16 = 4 \times 4$
1+2+3+4+**5**+4+3+2+1=	$25 = 5 \times 5$
1+2+3+4+5+**6**+5+4+3+2+1=	$36 = 6 \times 6$
1+2+3+4+5+6+**7**+6+5+4+3+2+1=	$49 = 7 \times 7$

CHILD'S PLAY?

Sox Unseen

Sam's favorite colors are blue and green, so it's not surprising that he has six blue sox and six green sox in his sock drawer. Unfortunately, they are hopelessly mixed up and one day, in complete darkness, he has to grab some sox to wear.

How many sox does he have to take from the drawer to make sure he gets a matching pair—either green or blue? (For some strange reason, his mother insists that his socks have to match!)

Answer, page 217

Gloves Galore!

Gloria's favorite colors are pink and yellow. She has sox in those colors, of course, but she *really* likes gloves!

In her glove drawer, there are six pairs of pink gloves and six pairs of yellow gloves, but like Sam's sox, the gloves are all mixed up. In complete darkness, how many gloves does Gloria have to take from the drawer in order to be sure she gets one pair? She doesn't mind whether it's a pink or yellow pair.

Answer, page 217

Hint: This may sound a bit like "Sox Unseen," but watch out! Gloves are more complicated than sox.

Birthday Hugs

"O frabjous day! Calloo Callay!"*
                    It's Jenny's birthday!

Jenny invites her three best friends Janey, Jeannie, and Joany to come to a party at her house, and when they all arrive they all give each other hugs.

How many hugs is that altogether?

*A special "gold star" if you can name the
 work and author of this famous line.

Answer, page 217

MATHBIT

The last joint of your thumb is probably close to an inch long, measuring from nail to knuckle. The spread from the tip of your thumb to the tip of your forefinger is probably five or six inches. Measure them with a ruler—then you can use these "units" to measure the lengths of all sorts of things.

Sticky Shakes

For John's birthday celebration, he invites six friends—Jack, Jake, Jim, Joe, Julian, and Justin—to a favorite burger place where they order thick and sticky milk shakes: banana, chocolate, maple, peanut butter, pineapple, strawberry, and vanilla.

While they slurp the shakes, their hands get sticky. Laughing about "shake" hands, they decide to actually do it—shake hands with their shake-sticky hands. So each boy shakes hands once with everyone else. How many handshakes is that altogether?

Answer, page 218

The Wolf, the Goat, and the Cabbage

You are traveling through difficult country, taking with you a wolf, a goat, and a cabbage. All during the trip the wolf wants to eat the goat, and the goat wants to eat the cabbage, and you have to be careful to prevent either calamity.

You come to a river and find a boat which can take you across, but it's so small that you can take only one passenger at a time—either the wolf, or the goat, or the cabbage.

You must never leave the wolf alone with the goat, nor the goat alone with the cabbage.

So how can you get them all across the river?

Answer, page 218

Floating Family

Mom and Dad and two kids have to cross a river, and they find a boat, but it is so small it can carry only one adult or two kids. Luckily both the kids are good rowers, but how can the whole family get across the river?

Answer, page 218

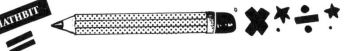

Did you know that most drinking glasses and cups have a circumference greater than their height? Test it out on some you have at home.

Take a piece of string and wrap it carefully once around a glass. You will almost always find the string is longer than the height of the glass. When is this not true?

Now you can amaze your friends by predicting this fun mathematical fact with one of their glasses before you measure it!

Slippery Slopes

Brenda the Brave sets off to climb a mountain which is 12,000 feet high. She plans to climb 3000 feet each day, before taking overnight rests. A mischievous mountain spirit, however, decides to test Brenda's resolve. Each night, Brenda's sleeping bag, with her soundly asleep in it, is magically moved 2000 feet *back down* the mountain, so that when Brenda awakes in the morning she finds herself only 1000 feet higher than she was the morning before!

Not one to give up, Brenda eventually succeeds. But how many days does it take her to reach the summit?

Answer, page 218

The Long and the Short of the Grass

Mr. Greengrass wants his lawn to be tidy and likes the grass cut short. Because he doesn't like mowing but wants to be able sit outside and read the paper on Sunday mornings and be proud of the smooth lawn, he decides to hire some good young mowers.

Two kids agree to mow Mr. Greengrass's grass on Saturdays for 15 weeks. To make sure they come every single Saturday, he agrees to pay them, at the end of the 15 weeks, $2 for every week that they mow it—as long as they will give him $3 for every week they miss.

At the end of the 15 weeks, they owe him exactly as much as he owes them, which is good news for Mr. Greengrass, but a rotten deal for the kids! How many weeks did they miss?

Answer, page 218

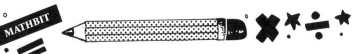

When drawing a graph, some people can never remember which is the x-axis and which is the y-axis. Here's a neat way to remember: say to yourself, "x is *a cross*."

Nine Coins

Wendy got into trouble in her math class. She was sorting out money she planned to spend after school, and accidentally dropped nine coins on the floor. They fell with such a clatter that the teacher was angry at the disturbance and told Wendy to remain at her desk after school until she could arrange all nine coins on the desktop in at least six rows with three coins in each row.

Can you do it?

Wendy did. In fact she did even better. She arranged her nine coins in no less than *ten* rows, with three in each row! Her teacher was quite impressed.

Can you make ten rows?

Answer, page 218

Remember pi to five decimal places by counting the number of letters in each word in the following question:

CAN I FIND A SMART PINEAPPLE?   (Pi = 3.14159)

Tricky Connections

Three new houses (below) have been built along a highway in Alaska. Each house needs an electricity supply and a water supply; however the permafrost means nothing can be buried underground, and no supply must ever cross a driveway. Too, a new safety law states that no electricity supply may cross a water pipeline.

Can the houses be connected up?

Answer, page 219

Odd Balls

Lucky you! You have nine tennis balls...

...and four shopping bags.

Your challenge is to put all the balls in the bags in such a way that there is an odd number of balls in each bag. That is, each bag must contain 1, 3, 5, 7, or 9 balls.

Can it be done?

Answer, page 219

Hint: Yes, it can, but there's a trick to it!

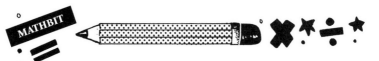

Challenge your friends to write down the largest possible number using only two digits. They'll probably write 99, but then you can top them.

The correct answer is 9^9, which means $9\times9\times9\times9\times9\times9\times9\times9\times9$, or 387,420,489 (nearly four hundred million)!

Cube of Cheese

Honoria was hosting a party—entertaining some friends. She had planned a specially elegant dinner, and wanted a cube of cheese as part of an appetizer display.

Looking in the refrigerator, she found she did have some cheese, but it was in the form of a complete sphere; in other words, a ball of cheese. Well, she would simply have to cut a cube of cheese from the cheese ball.

Honoria can't resist a puzzle so she spent most of the time as she prepared dinner wondering how to cut the sphere of cheese into a cube quickly; that is, using the fewest number of cuts.

What is the smallest number of cuts you have to make to cut a cube from a sphere?

Answer, page 219

Potato Pairs

In Idaho, they proudly say they have giant potatoes, and unusual potato sellers. One of the strangest potato sellers is Potato Mo. She never sells her potatoes one at a time, nor in bags of five or ten pounds. She sells potatoes only in pairs!

One day, Cal the cook wanted a potato that weighed just two pounds, so he went and asked Mo what she had available.

"I have only three potatoes left," she answered. "Here they are: A, B, and C.

"A and B together weigh three pounds; A and C together weigh five pounds; B and C together weigh four pounds. You can have any pair you want."

Can you help Cal the cook buy a pair of potatoes? Which if any of the potatoes weighs two pounds?

Answer, page 219

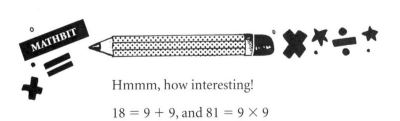

Hmmm, how interesting!

$18 = 9 + 9$, and $81 = 9 \times 9$

Sugar Cubes

The Big Sugar Corporation wants to persuade people to use lumps of sugar, or sugar cubes; so they run a puzzle competition. The first person to get the answers right (the puzzle is made up of three parts) wins free sugar for life! Here's the puzzle:

You have been sent a *million cubes* of sugar. Yes, that's right, 1,000,000 sugar cubes! Each cube is just half an inch long, half an inch wide, and half an inch high.

1. Suppose the cubes arrived all wrapped up and packed together into one giant cube. Where could you put it? Under a table? In the garage? Or would you need a warehouse? (*Hint: What you need to work out is, how many little cubes would there be in each direction? And how long, wide and high would the giant cube be?*)

2. Now, suppose you decide to lay the cubes all out in a square on the ground—all packed together but this time only one layer deep? How big a space would you need? Your living room floor? A tennis court? Or would you need a parking lot the size of a city block?

3. Now for the big one. Pile all the million cubes one on top of the other into a tower just one cube thick. (You'll need very steady hands and not a breath of wind!) How high will the pile be? As high as a house (say 25 feet)? As high as New York's Empire State Building (1472 feet)? As high as Mount Adams (12,000 feet) in Washington state or Mount Everest (29,000 feet)? Or will the pile of cubes reach the moon (240,000 miles)?

Answer, page 220

Crackers!

Mad Marty, crazy as crackers, invites his friends to a cracker puzzle party. The puzzle he sets them is this: How many different kinds of spread can you put on a cracker?

Everyone brings a different kind of spread and Marty supplies a gigantic box of crackers. Then they all get down to business:

Marty has a cracker with mayo =   1 spread

Pete brings peanut butter; so now
they have: (**1**) mayo, (**2**) peanut
butter, (**3**) mayo and peanut
butter =   3 spreads

Jake brings jelly; so now they have
   (**1**) mayo, (**2**) peanut butter,
   (**3**) mayo and peanut butter,
   (**4**) jelly, (**5**) jelly and mayo,
   (**6**) jelly and peanut butter,
   (**7**) jelly and mayo and peanut
butter =   7 spreads

Hank brings honey =   how many
spreads?

Charlie brings cheese =   how many
spreads?

Fred brings fish-paste =   how many
spreads?

Answer, page 220

Crate Expectations

You have six bottles of pop for a party, and you want to arrange them in an attractive pattern in the crate. Four will make a square...

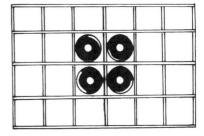

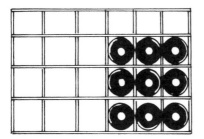

and nine will make a square. But six is a trickier number. How about an even number of bottles in each line?

Can you arrange them so that, in every row and in every column, the number of bottles is even (0, 2, 4, or 6)?

Answer, page 221

Hint: This is quite tricky, and a fine puzzle to challenge your friends with. A good way to practice is to draw a grid on a piece of paper and use coins instead of bottles.

Witches' Brew

Three witches were mixing up a dreadful mathematical spell in their cauldron, and one of them—Fat Freddy—was reading out the recipe to the others.

Eye of newt and toe of frog
Wool of bat and tongue of dog

Suddenly they realized they needed some liquid—2 pints of armpit sweat. They had a bucketful of sweat, a saucepan that when full held exactly 3 pints, and a jug that when full held exactly 1 pint. How could they get exactly 2 pints?
Answer, page 221

Hint: Try filling the pan, and then filling the jug from it.

Witches' Stew

Many years later the same witches, now even older and more haggard, were mixing up a super-disgusting stew in their cauldron:

Adder's fork and blind worm's sting
Lizard's leg and howlet's wing...

And once again they needed to add the sweat, mixed this time with tears. They had a bucketful of liquid, and they needed to add exactly 4 pints, but all they had to measure it was a pitcher that held exactly 5 pints and a pot that held exactly 3.

How could they measure out exactly 4 pints?

Answer, page 221

The Pizza and the Sword

The room is full of hungry people. You have just had delivered a monster pizza, which covers most of the table. It's too hot to touch, but you need to cut it up quickly so that everyone can start eating.

What is the maximum number of pieces you can make with three straight cuts across? You may not move the pieces until you have finished cutting; so you can't pile them on top of one another!

You could make the three cuts side by side, which would give you one extra piece for each cut; so you would get four pieces altogether.

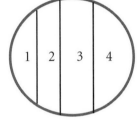

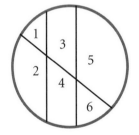

Or, you could make two cuts side by side, and then the third across both of them. That would give you six pieces. But could you make more than six.
Answer, page 221

Hint: Nobody said the pieces all have to be the same size; indeed it would be better if they were all different sizes!

Pencil Squares

Lay 15 pencils (or straws, toothpicks or what have you) out on the table to make five equal squares like this:

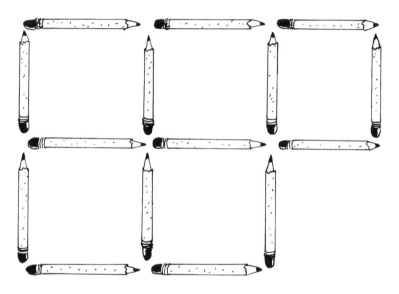

Now take away just three of the pencils, and leave only three squares.

Answer, page 222

Pencil Triangles

Here's a really tough puzzle that you can use to stump your friends, after you have figured it out. If you are able to solve it without peeking at the solution, you are doing better than a whole lot of brilliant people.

This time, take 6 pencils (or straws or toothpicks) and arrange them so that they form four equal triangles.

Answer, page 222

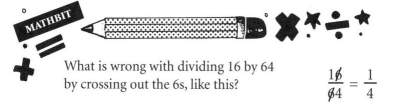

What is wrong with dividing 16 by 64 by crossing out the 6s, like this?

$$\frac{1\cancel{6}}{\cancel{6}4} = \frac{1}{4}$$

Cancelling the 6s in the division 16/64 is wrong because each is only part of a number, and you must always cancel a whole number. Although it seems to work in this case, it's just by luck. (You get into terrible trouble if you cancel, say, the 5s in 15/45; you get 1/4, but the right answer is 1/3!)

Cyclomania

In the kids' playground, Donna was delighted to find bicycles with two wheels, and tricycles with, of course, three wheels.

They came in all sorts of different shapes and sizes and colors, but she took a count and discovered that they had 12 wheels altogether.

How many bicycles did Donna find there? And how many tricycles?

Answer, page 222

Here's another puzzle question to stump your pals. Can you make 100 using just four 9s? How?

Answer, page 232

Spring Flowers

On her breakfast tray, Aunt Lily had a little vase of flowers—a mixture of primroses and celandines. She counted up the petals and found there were 39. "Oh, how lovely!" she said, "exactly my age; and the total number of flowers is exactly your age, Rose!"

How old is Rose?

Answer, page 223

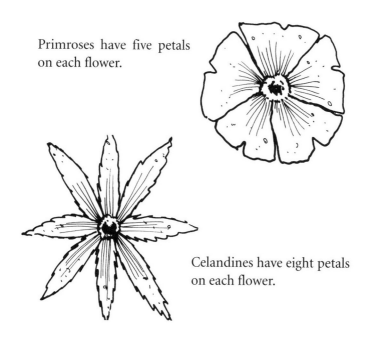

Primroses have five petals on each flower.

Celandines have eight petals on each flower.

Sesquipedalian Farm

On an ordinary farm, if one goose lays one egg in one day, then you can easily work out how many eggs seven geese lay in a week, can't you? One goose lays seven eggs in a week; so seven geese lay seven times that many—that's 49 eggs a week.

But, on Sesquipedalian Farm things are a bit different, because a hen and a half lays an egg and a half in a day and a half (how very strange!). So how many eggs would seven hens lay in a week and a half?

Answer, page 223

Three-Quarters Ranch

It's no surprise that calculations get even more complicated at Three-Quarters Ranch, where a duck and three quarters lays an egg and three-quarters in a day and three-quarters.

How many eggs do seven ducks lay in a week?

Answer, page 223

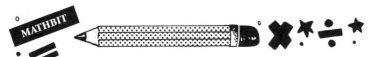

You will need a calculator for this trick. First secretly enter 271 and × or * to multiply. Then ask your friend what her favorite digit is, from 2 to 9. While she watches, punch it in, and press =. Then give her the calculator, and tell her to multiply by 41, and see what she gets; all her own digits! To find out how this works, multiply 41 × 271 and see if you can figure it out.

Cookie Jars

Joe and Ken each held a cookie jar and had a look inside them to see how many cookies were left.

Joe said, "If you gave me one of yours, we'd both have the same number of cookies."

Ken said to Joe, "Yes, but you've eaten all yours, and you haven't any left!"

How many cookies does Ken have?

Answer, page 224

The equals sign = was invented in 1557 by Welsh mathematician Robert Recorde. In his book, *The Whetstone of Witte,* he wrote, "To avoid the tedious repetition of these words: is equal to; I will set a pair of parallel lines thus, =, because no two things can be more equal."

Fleabags

Two shaggy old dogs were walking down the street.

Captain sits down and scratches his ear, then turns to Champ and growls, "If one of your fleas jumped onto me, we'd have the same number."

Champ barks back, "But if one of yours jumped onto me, I'd have five times as many as you!"

How many fleas are there on Champ?

Answer, page 224

The Rolling Quarter

Imagine a quarter laid on the table and fixed there, perhaps with a dab of glue. Now lay another quarter against it, and roll the second quarter all the way around the first one, without any slipping at the edges, until it gets back to where it started.

In making one complete circuit, how many times does the second quarter rotate?
Answer, page 224

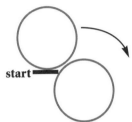

start

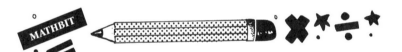

Have you seen the patterns in the sevenths, when turned into decimals? Use a calculator to work out 1/7, 2/7, 3/7, and so on, and this is what you get:

1/7 =	0.14285714	285714	2857...
2/7 =	0.285714	285714	285714...
3/7 =	0.4285714	285714	28571...
4/7 =	0.57142	857142	8571428...
5/7 =	0.7142	857142	85714285...
6/7 =	0.857142	857142	857142...

Sliding Quarters

Here's a puzzle that looks simple, but is really quite tricky. Even when you have seen the answer you sometimes can't remember it. Maybe you are smart enough to solve it on your own, but if you go on to baffle your friends, make sure you can remember the solution when you need it!

Lay six quarters on the table touching in two rows of three, like this:

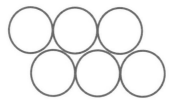

In each move, slide one quarter, without moving any others, until it just touches two others. In only three moves, can you get them into a circle like this?

Answer, page 224

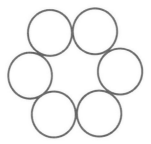

Picnic Mystery

Allie takes fruit, cake, and cookies for her picnic. She has three boxes for them. One is labeled FRUIT. One is labeled COOKIES. One is labeled CAKE. But she knows her Mom likes to fool her and has put every single thing in the wrong box. The only other thing she knows for sure is that the fruit is not in the CAKE box.

Where is the cake?

Answer, page 225

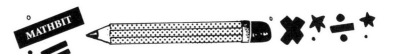

What if the combination lock you have has only three digits, and they only go from 0 to 6? Then the total number of possible combinations is only $6 \times 6 \times 6$ or 216, and a thief could very likely go through all those numbers to open the lock in less than four minutes!

Find the Gold

Lucy Sly, a brilliant detective, has tracked some pirates to their island base. In their secret cave, she finds the pirate chief with three chests of treasure. One chest contains pieces of iron, one chest pieces of gold, and the third has a mixture.

In return for a chance to escape, the pirate chief offers Lucy one chest to take away with her. All three chests are labeled—IRON, GOLD, and MIXTURE. But, he warns her, all the labels are on the wrong chests.

"Then I can't tell which is which," she replies.

"I will take one object out of any one of these chests, and show it to you—although you may not look inside."

Which chest would you choose to see an object from? And how would you be sure you got the chest of gold?

Answer, page 225

Frisky Frogs

Across a stream runs a row of seven stepping stones.

On one side of the stream, sitting on the first three stones, are three girl frogs, Fergie, Francine, and Freda, and they want to get across to the other side.

There's an empty stone in the middle.

On the other side are three boy frogs, waiting to come across the other way—Fred, Frank, and Frambo.

Only one frog moves at a time. Any frog may hop to the next stone if it is empty, or may hop over one frog of the opposite sex on to an empty stone.

Can you get all the frogs across the river?

Answer, page 225

Leaping Lizards

Across a stream runs a row of eight stepping stones.

On one side of the stream, on the first five stones, sit five girl lizards—Liza, Lizzie, Lottie, Lola, and Liz—and they want to get across to the other side.

There's one empty stone in the middle.

On the other side are three boy lizards, waiting to come across the other way—Lonnie, Leo, and Len.

Only one lizard moves at a time. Any lizard may hop to the next stone if it is empty, or may hop over one lizard of the opposite sex onto an empty stone.

Can you get all the lizards across the river, and what's the smallest number of leaps?

Answer, page 225

Many spiders weave beautiful roundish webs, with a single strand spiraling out from the center. These amazing creatures keep the distances and turns so exact. Watch for webs on damp and frosty mornings and count the radial lines used in their construction.

Chewed Calculator

You look for your calculator to help work out some figures, and when you finally find it you can see that the dog has been chewing on it! He has chewed up all the number buttons, so that not one of them works, except for the 4. What's more, the 4 button only seems to work if you press it four times, and then you get four 4s!

All of the other calculator keys work: (*) (/) (+) (−) (=) (sqrt) (MS) (MR) and (1/x), but strangely enough no other numbers.

How can you get all the numbers from 1 to 10 using only four 4s for each? *Answer, page 000*

Examples:

You can make 1 by punching in 4/4=*4/4=

You can make 2 by punching in 4+4=MS4*4=/MR=

Can you figure out how to make 3, 4, 5, and the other numbers up to 10? *Answer, page 225*

Crushed Calculator

This time, you are shocked to find out that a small tame elephant has sat on your calculator! When you try to use it, you find that all the function buttons are working: (+) (−) (/) (*) (=) (sqrt) (1/x).

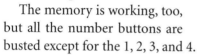

The memory is working, too, but all the number buttons are busted except for the 1, 2, 3, and 4.

Using only the working 1, 2, 3, and 4 buttons and using each of them once every time, can you make all the numbers from 1 to 20? *Answer, page 000*

Examples:

You can make 5 by punching 4*2=*1=-3=
You can make 6 by punching 4/2=*3=*1=

Are you able to work out the others? You need to be clever for this one. *Answer, page 226*

Squares & Cubes

Multiply any number by itself, and you get a square number. So $2 \times 2 = 4$, and 4 is a square number. And four squares fit together to make a bigger square.

Nine is also a square, because $3 \times 3 = 9$, and nine squares also fit together to make a bigger square.

Cubes are numbers you get by multiplying a number by itself and then by the same again; so $2 \times 2 \times 2 = 8$, which is a cube.

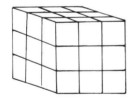

And $3 \times 3 \times 3 = 27$, another cube. The big cube is made of 27 little cubes.

There is only one 2-digit number (i.e., between 10 and 99) that is both a square and a cube. What is it?

Answer, page 226

Cubes & Squares

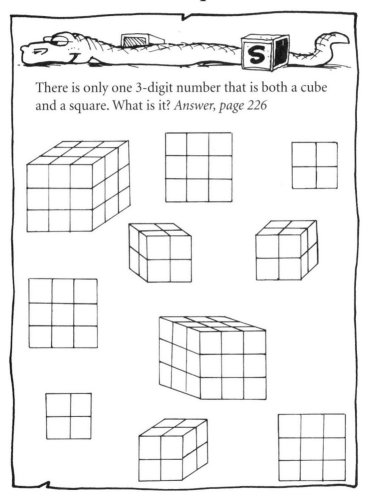

There is only one 3-digit number that is both a cube and a square. What is it? *Answer, page 226*

Old MacDonald

Old MacDonald had a farm, EE-I-EE-I-OH!
And on that farm he had some pigs, EE-I-EE-I-OH!
With an *Oink oink!* here, and an *Oink oink!* there.
Here an *Oink!* There an *Oink!*
Everywhere an *Oink oink!*
Old MacDonald had a farm, EE-I-EE-I-OH!

Old MacDonald had some turkeys, too (certainly with a *Gobble gobble* here and a *Gobble gobble* there).

One day, while out feeding them all, he noticed that, if he added everything together, his pigs and his turkeys had a total of 24 legs and 12 wings between them.

How many pigs did Old MacDonald have? And how many turkeys? *Answer, page 226*

Old Mrs. MacDonald

Mrs. MacDonald was a farmer too. She kept the cows and chickens. One day when she went out to feed them she counted everything up, and found that her animals had a total of 12 heads and 34 legs.

How many cows did she have? How many chickens?

Answer, page 226

If you use a combination lock, you can easily work out how long it will take a thief to try all the numbers and open it. If it has four dials with 10 digits on each, then there are a total of 10,000 different combinations. If the thief takes one second to try each, it will take nearly three hours to go through every number, since in three hours there are $3 \times 60 \times 60$ seconds, or 10,800 seconds. On average, though, a thief will reach your secret number in half that time—say an hour and a half. (See Mathbit on page 50.)

Wiener Triangles

In the link-wiener factory in Sausageville, the wieners are made in long strings, with a link of skin holding each sausage to the next one. So, although the wieners are firm, you can bend the string of wieners around into many shapes. For example, you can easily bend a string of three wieners into a triangle.

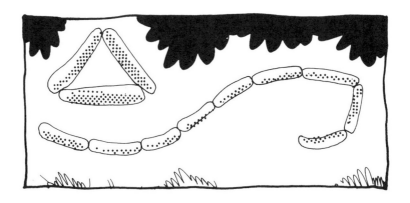

Suppose you had a string of 9 wieners. Without breaking the string, how many triangles can you make?

Answer, page 227

Tennis Tournament

You successfully arranged a "knock-out" tennis tournament, in which the winners of the first round meet in the second round, and so on. The little tournament had only four players, so arranging it was easy.

In the first round, Eenie played Meanie, and Eenie won. Miney played Mo, and Mo went through to the second round. In the second round—the final—Mo beat Eenie, and won the tournament.

The 3-game match card looked like this:

The match was so well organized, you've been asked to arrange another knock-out tennis tournament. This time 27 players enter. How many matches will have to be played to find the winner?

Answer, page 227

House Colors

George and Greg Green both live with their families in big houses on Route 1, just outside town, and their sister Bernice, married to Bert Blue, lives in the third house on the same road.

One day, they all secretly decided to paint their houses the same color as their names, and were glad to find out afterwards that no next-door houses were the same color.

Who lives in the middle house of the three?

Answer, page 227

MATHBIT

How many people do you think need to be in a room before there is a fifty/fifty chance that two of them will have the same birthday? *Answer, page 217*

LoadsaLegs

Two multipedes were dancing together at a party, and trying hard not to trip over each other's feet! One smiled at the other and said, "If you could give me two of your legs, we'd have the same number." The other replied, "If I had two of yours, I'd have three times as many legs as you!"

How many legs did each have?

Answer, page 227

Antennas

Two robots are strutting through cyberspace, trying to keep in touch with everything on the World Wide Web.

Abot says to Bbot, "I get such a headache trying to listen to all these radio signals at the same time. Do you know that if you gave me two of your antennas we'd have the same number?"

Bbot retorted, "That's nothing! If I had two of your antennas I'd have five times as many as you!"

How many antennas does each of them have?

Answer, page 227

All numbers have factors, which divide evenly into them. The factors of 8 are 1, 2, and 4.

Perfect numbers are special because they are equal to the sum of their factors. The number 6 is perfect because its factors are 1, 2, and 3, and their sum $1 + 2 + 3 = 6$. The next perfect number is 28, because it is the sum of 1, 2, 4, 7, and 14. The third perfect number is 496!

The Power of Seven

Far back in history, a lonely fort was being desperately defended against thousands of attackers.

The attacks came regularly at noon every day, and the defending commander knew he had to survive only three more days, for then would come the end of the attackers' calendar, and they would all go home to celebrate, giving time for his reinforcements to arrive.

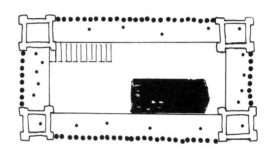

He also knew that the attackers held an unshakable belief in the power of the number seven. So he always placed seven defenders on each wall of the fort. With three attacks to come, and only 24 defenders, he places 5 along each wall, and 1 in each corner tower.

The attackers charge in from the north, and see seven defenders along that wall. Firing a volley of arrows, they wheel round and retreat, chanting "Neves! Neves!" meaning seven in their language. They charge from the west and again see seven defenders facing them. Firing a volley of arrows they retreat again. "Neves! Neves!"

From the south, then the east, again they charge. Each time they are met by exactly seven defenders. Each time they turn and flee, chanting "Neves! Neves!" And the attack is over for the day.

The commander mops his worried brow as the bugler blows the bugle to signal "Well done and all clear!" Then he learns the arrows have killed four of his men.

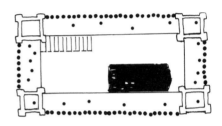

How can he rearrange the remaining 20 so that by noon of the next day there will still be 7 defenders on each side?

Answer, page 228

The Power of Seven Continues

At noon on the second day the pattern of attack is different; the attackers come from the west, from the south, from the north, and then from the east. Each time they see seven defenders, fire a volley of arrows, and retreat, chanting "Neves! Neves!"

The attack is over, but five more men have been killed. Is it still possible for the commander to place seven defenders along each wall, now that he has only 15 altogether?

In other words, can they survive that third, final day of attack?

Answer, page 228

Bundles of Tubes

William Posters set up a company to make cardboard tubes and sell them for the protection of large pictures or posters sent through the mail. He advertised:

KEEP YOUR POSTERS SAFE WITH BILL POSTERS' TUBES!

He makes some tubes 2 inches in diameter and some 3 inches in diameter, for extra big posters. He sells the bigger tubes in bundles of 19, and the smaller tubes in bundles of 37.

Most people sell things in tens, or dozens, or even packets of 25. Why do you think he chose 19 and 37?

Answer, page 228

Pyramids

Susie and Ben like to make quick and easy cannonball cookies, so they often make lots and lots.

Today, they decide to heap the cookies up on the table for the family in pyramid shape.

Susie decides to make a triangular base with 6 cookies along each side, and builds up her pyramid from there—5 along each side in the next layer, then 4, then 3, and so on up to 1 on top.

Ben starts by laying out a square on the table, with only 5 cookies on each side. Then he builds up 4 on each side, then 3, and so on.

Which of the two pyramid builders uses more cookies by the time they reach the top?

Answer, page 229

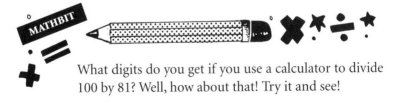

What digits do you get if you use a calculator to divide 100 by 81? Well, how about that! Try it and see!

Wrong Envelope?

You decide one day to write a letter to each of three friends. When you finish, you go to a desk and find three envelopes, write the address on each envelope, and stick on a stamp.

Now, suppose you were to put one letter into each of the envelopes without looking at the front of it, how many ways are there of putting at least one letter into the wrong envelope?

What is the chance that you will get the letters in the right envelopes just by luck? *Answer, page 229*

Squares & Cubes & Squares

1) Which 2-digit number is 1 more than a square, and 1 less than a cube? *Answer, page 229*

2) Which three-digit number, made of consecutive digits, like 567, is 2 less than a cube and 2 more than a square?
Answer, page 229

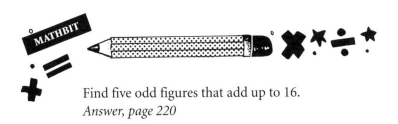

Find five odd figures that add up to 16.
Answer, page 220

Good Neighbor Policy

Kind old Mrs. Werbenuik always liked to present her friends with special gifts during the holidays. One year she took pottery classes on Wednesday evenings and, while there, she made three fancy little pots. Later, she filled them with special homemade apple jelly, and gave them to her neighbors—a family of two fathers and two sons. They were very happy with their gifts.

How could the three pots of jelly be divided equally and fairly between two fathers and two sons?
Answer, page 230

Cutting the Horseshoe

Here's a picture of a horseshoe.

Your challenge is to cut it into seven pieces, each containing a nail hole, with just two straight cuts. After the first cut, you can put the pieces on top of one another, but both cuts have to be straight.

Can you make seven "holey" pieces?

Answer, page 230

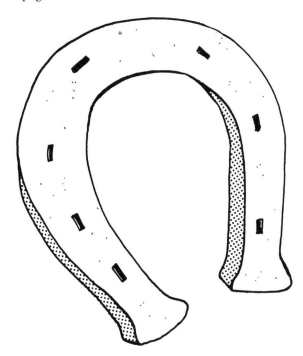

Multisox

Two multipedes are cantering through a shopping mall when they come to a sock shop. They count up all their feet and find that three dozen, or 36, socks will be just enough to keep all of their feet warm.

If one of them has eight feet more than the other, how many feet does each multipede have?

Answer, page 230

Three Js

Joan and Jane are sisters. Jean is Joan's daughter, and 12 years younger than her aunt.

Joan is twice as old as Jean.

Four years ago, Joan was the same age as Jane is now, and Jane was twice as old as her niece.

How old is Jean?

Answer, page 230

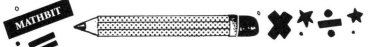

You can easily find if a big whole number can be divided by 3. Just add up the digits, and go on adding until you get only one digit. If that digit is a 3, 6, or 9, the larger number is divisible by 3; if not, not. So, try 12; 1 + 2 = 3; 12 is divisible by 3.

256 2 + 5 + 6 = 13; 1 + 3 = 4

256 is *not* divisible by 3

5846 5+8+4+6 = 23; 2+3=5

5846 is *not* divisible by 3

7293654 7+2+9+3+6+5+4 = 36; 3+6=9

7293654 *is* divisible by 3

Architect Art

Art became an architect,
> And thought he'd like to draw
>> A house in three dimensions,
>>> But was not entirely sure
>>>> That he could do it in one go,
>>>>> And never lift his pen
>>>>>> Above the page, and not go o'er
>>>>>>> Any line again.

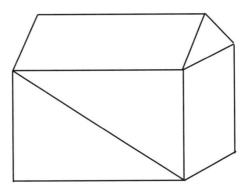

Can you do it? Can you draw this house with one continuous line, without lifting your pencil from the paper or going over any line twice? *Answer, page 230*

Hint: Start from a corner where an odd number of lines meet.

No Burglars!

Worried by the number of burglaries in your town, you have just installed Fantastico High-Security Locks on all the doors in your house. They are so special that you actually have to walk through a doorway and lock the door behind you, and then it cannot be opened by anyone else.

Here is a plan of your house:

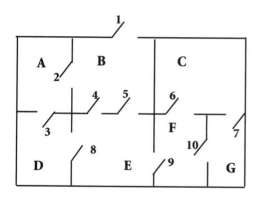

A = bedroom
B = hall
C = dining room
D = bathroom
E = living room
F = kitchen
G = den

You decide to go out to the movies. You need to go through and lock each door, ending with the front door.

In which room would you start? And in which order would you shut the doors behind you to make sure you went through and locked every one?

Answer, page 231

Train Crash

There's a single railroad track across the remote desert near the Arizona–New Mexico border. A freight train starts from one end and goes north at 25 mph. An ancient pioneer train starts at the other end and coughs its way south at a mighty 15 mph.

Neither driver sees the other train approaching, and at No Hope Gulch, after both trains have been traveling for exactly one hour, they collide head-on.

There's a lot of arguing about who's to blame, but the question is, how far apart were the trains when they started, exactly one hour before the crash?

Answer, page 231

Puzzle of the Sphinx

Here are four little sphinxes, small versions of the big one found in the Egyptian desert.

How can you rearrange these four sphinx shapes to make one big sphinx? *Answer, page 231*

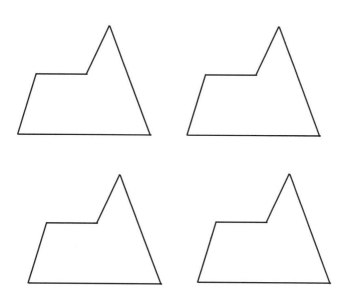

Hint: It's really just a small jigsaw puzzle, except that one of the "pieces" fits in backwards.

Perforation!

You have 12 postage stamps that have a picture of your favorite flower on each.

You want to put them in your stamp book, but it's made for 3 rows of 4 stamps like this, not 4 rows of 3.

How can you tear the sheet of stamps, along the perforations,

into only two pieces so that they will fit together and fill your page better? *Answer, page 231*

Disappearing Apples

Joe bought a bag of apples on Monday, and ate a third of them. On Tuesday he ate half of the remaining apples. On Wednesday he looked in his bag and found he had only two apples left.

How many did he have to start with?

Answer, page 232

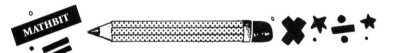

In the late sixteenth century, the Italian scientist Galileo was gazing at a church lamp swinging to and fro and realized the time of each swing was always the same. Making pendulums by tying weights on various lengths of string, he discovered that when he doubled the string length, the swing of the pendulum took four times as long. How long would a pendulum take to swing if its string were three times as long?

Colored Balls #1

You have five red balls,

five yellow balls,

and five blue balls.

How can you arrange them in this triangular frame so that no two balls of the same color are next to one another? *Answer, page 232*

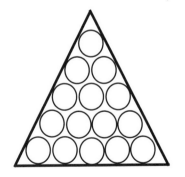

Colored Balls #2

You have six yellow balls

and four blue balls.

How can you arrange them in this triangular frame so that no three yellow balls make an equilateral triangle—a triangle in which all three sides are the same length?

Answer, page 232

Logical Pop

You have three identical cans. One is labeled POP, one is labeled MILK SHAKE, and one is labeled POTATO CHIPS.

However, you know your mischievous sister has, as a joke, changed all the labels, so that every label is on the wrong can.

You want to open the pop, but you are allowed to open only one can; so you have to get it right.

To get a clue, you may shake one can first. Which one would you shake, and how would you choose which can to open?

Answer, page 232

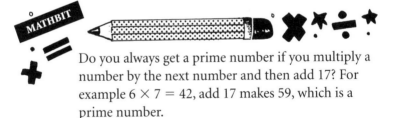

Do you always get a prime number if you multiply a number by the next number and then add 17? For example 6 × 7 = 42, add 17 makes 59, which is a prime number.

WORKING TOWARDS WIZARDRY

Secret Number Codes

To send a secret message to a friend, send the message in code so that no one else can read what it says. Of course, your friend needs to know how to decode the message, so only the two of you will understand.

This simple code changes each letter to a number:

A	B	C	D	E	F	G	H	I	J	K	L	M
1	2	3	4	5	6	7	8	9	10	11	12	13

N	O	P	Q	R	S	T	U	V	W	X	Y	Z
14	15	16	17	18	19	20	21	22	23	24	25	26

To send the message MEET ME AT SIX you write down their numbers instead of the letters:

13 5 5 20   13 5   1 20   19 9 24

But clever people might be able to guess this code, so you might want to make yours a little more complicated. See if you can work out what this message means:

14 6 6 21   14 6   2 21   20 6 23 6 15

Answer, page 233

Hint: This second code is just a bit harder than the one above.

Shape Code

This code looks much more mysterious. By using a shape instead of a number to stand for each letter, the same message—MEET ME AT SIX—looks so weird you might think it has no meaning at all. It doesn't even look like any kind of language, yet it's a marvelous but simple code. Here's how to use it.

Make a grid and write in the letters as shown:

AB	CD	EF
GH	IJ	KL
MN	OP	QR

Each space stands for one of the letters that was in it. The first letter of the two is shown by just the space; the second letter is shown by the space with a dot in it.

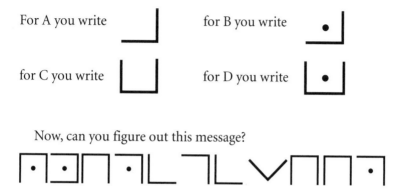

For A you write   for B you write

for C you write   for D you write

Now, can you figure out this message?

Answer, page 233

Magic Triangle

Here's a triangle of ovals:

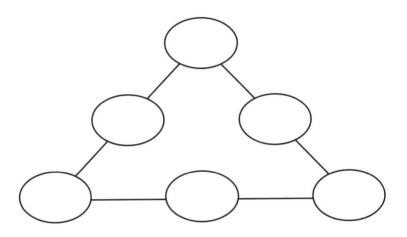

Can you write the numbers 1 through 6 in the circles, using each number once, so that the total you get by adding the numbers along each side always comes to 9?

Answer, page 233

Magic Hexagon

Here is a section of a honeycomb—seven hexagons in a group.

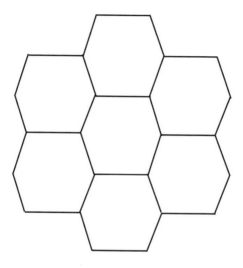

Can you write the numbers 1 through 7, one in each hexagon, so that all three lines across the middle add up to a total of 12?

Answer, page 234

Easier by the Dozen

Place the numbers from 1 to 12 as follows:

The odd numbers go in the triangle. The even numbers go in the circle. The numbers that are divisible by three go in the square.

How will this look? *Answer, page 234*

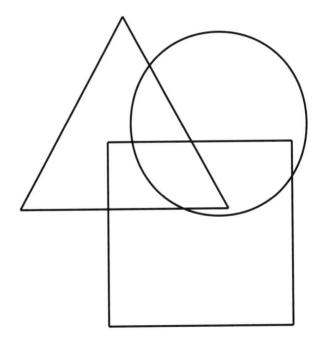

Hint: The puzzle would not be possible if the figures didn't overlap. Pay careful attention to the numbers that are multiples of 3—namely, 3, 6, 9, and 12.

Who Is Faster?

Hector can run from the train station to his parents' house in eight minutes. His younger brother Darius can run the same distance eight times in one hour. (Not that he'd need to!) Who is faster?
Answer, page 234

Hint: How long does it take Hector to run the distance eight times?

The Average Student

Melissa got a poor grade on her very first homework assignment at her new school—only one star out of a possible five stars! She was determined to do better. How many five-star ratings must she receive before she has an average rating of four stars?

Answer, page 235

Hint: Knowing that the average of the ratings must be four, you start out with a three-point difference between the one-star rating (given to Melissa on her first assignment) and the desired four-star average. Well, how many stars can you make up at a time?

Big Difference

Your challenge is to place the digits 2, 4, 6, and 7 into the boxes so that the difference between the two two-digit numbers is as big as possible. *Answer, page 235*

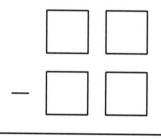

Not Such a Big Difference

What if you wanted to make the difference as *small* as possible? *Answer, page 235*

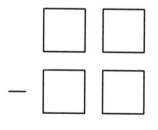

Hint (Big Difference): To find the biggest difference, try creating the biggest possible number.

Hint (Not Such a Big Difference): Find two numbers that are close together.

Square Route

Four dots are arranged in a square. Starting at the upper left dot, draw three straight lines, each line going through one or more dots, so that you end up where you started. Every dot should have a line going through it. *Answer, page 236*

● ●

● ●

The Missing Six

Place the six numbers below into the empty circles so that both sentences are true. Use each number once and only once.
Answer, page 236

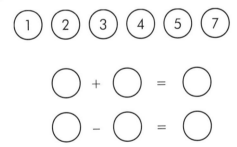

Hint (The Missing Six): More trial and error. If the "7" in the puzzle was replaced by a "6," no answer would be possible.

Hint (Square Route): All three lines go beyond the boundary of the square.

Donut Try This at Home

Suppose a low-calorie donut has 95 percent fewer calories than a regular donut. How many low-calorie donuts would you need to eat to take in as many calories as you'd get from a regular donut?
Answer, page 236

Hint: If a regular donut has, say, 100 calories, how many does a low-calorie donut have?

The Long Way Around

If the height of the diagram below is 8 units, and the length is 15 units, how far is it around the entire diagram?

Answer, page 237

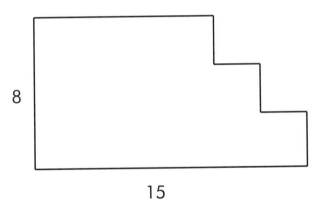

8

15

A Very Good Year

The year 1978 has an unusual property: When you add the 19 to the 78, you get 97—the middle two digits of the year!

What will be the next year to have this same property?

Answer, page 237

Hint: The year could not be in the 2000s or the 2100s, because the middle number would start off less than the first two digits. So start with the year 2200 and work from there. You need to do a little trial and error—but not too much!

Long Division

Professor Mathman went to the blackboard and demonstrated to his astonished class that one-half of eight was equal to three! What did the professor do?

Answer, page 237

Hint: Professor Mathman was being tricky. It's easier to do the problem if you think in terms of the number 8 rather than the word "eight."

When in Rome

Was the previous problem too easy? If so, try to come up with a way of proving that one-half of nine equals four.
Answer, page 237

Hint: Like problem #23, the answer to this one is visual in nature. Don't forget the title.

Diamond in the Rough

Of the four suits that make up a deck of cards, only the diamonds are symmetrical, in that a diamond—unlike a club, a heart, or a spade—looks the same whether it is rightside-up or upside-down.

However, one of the 13 diamond cards is different when you turn it upside-down. Without checking any decks of cards you may have lying around, can you name that one non-symmetrical diamond?

Answer, page 237

Hint: The idea is to visualize how the diamonds are put on the cards. The only hint you'll need is that there are never three diamonds in a row across the card, although for the higher numbers there are certainly three or more diamonds in a row going down the card.

Three's a Charm

There is an inexpensive item that can be purchased for less than a U.S. dollar. You could buy it with four standard U.S. coins. If you wanted to buy two of these items, you'd need at least six coins. However, if you bought three, you'd only need two coins. How much does the item cost?

Recall that you have only five U.S. coins to work with: A penny (one cent), a nickel (five cents), a dime (ten cents), a quarter (twenty-five cents), and a half-dollar (fifty cents).

Answer, page 238

Hint: It's probably easiest to look at the combinations of two coins. What combinations of two coins produce a number that is divisible by 3? For example, a quarter plus a penny equals 26 cents, which is not divisible by 3, so this combination can be ruled out. On the other hand, a nickel plus a penny equals 6 cents, which is divisible by 3 but isn't nearly big enough to satisfy the problem! Once you get the right combination of two coins, you can work backward to get the rest of the answer.

Who Is the Liar?

Four friends—Andrew, Barbara, Cindy, and Daniel—were shown a number. Here's what they had to say about that number:

> Andrew: It has two digits
> Barbara: It goes evenly into 150
> Cindy: It is not 150
> Daniel: It is divisible by 25

It turns out that one (and only one) of the four friends is lying. Which one is it? *Answer, page 238*

The Powers of Four

Bert and Ernie take turns multiplying numbers. First Bert chooses the number 4. Ernie multiplies it by 4 to get 16. Bert multiplies that by 4 to get 64. Ernie multiplies that by 4 to get 256.

After going back and forth several times, one of them comes up with the number 1,048,576. Who came up with that number, Bert or Ernie?

Don't worry—the problem is easier than it looks at first glance. You don't have to multiply the whole thing out to figure out the correct answer! *Answer, page 239*

Hint (Who Is the Liar?): First assume that Andrew is lying, and see if it is possible for Barbara, Cindy, and Daniel all to be telling the truth. Then do the same for the other three. In only one case will there be only one liar.

Hint (The Powers of Four): If you kept multiplying by 4, that would lead you to the right answer, but an easier approach is to look for patterns.

High-Speed Copying

If 4 copiers can process 400 sheets of paper in 4 hours, how long does it take 8 copiers to process 800 sheets?

Answer, page 239

COPYMATIC

Hint: Questions like this one have been around for a lot longer than there have been copiers! The best approach is to look at the number of copies an individual can make, and go on from there.

Divide and Conquer

Fill in the boxes below to make the division problem work out.
Answer, page 239

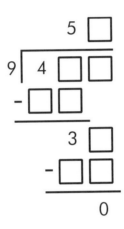

Agent 86

Fill in the missing squares in such a way that the rows, columns, and the two diagonals all add up to the same number.

Answer, page 239

32	19		8
10	25		
9			
35	16		11

Comic Relief

While traveling in Russia, I bought six comic books for a total of seventeen rubles. Some of the comics cost one ruble, others cost two rubles, while the most expensive ones sold for ten rubles apiece.

How many of each type did I buy?

Answer, page 240

Hint (Comic Relief): How many ten-ruble comic books can there be?

Hint (Agent 86): If you add up the numbers in the first column, you will find out the sum for every row, column, or diagonal. Then go on to those rows or columns containing three out of the four possible numbers, and you'll be able to figure out the fourth. Pretty soon you'll be all done!

Pieces of Eight

An octagon is an eight-sided figure. A stop sign is perhaps the most familiar example of a "regular" octagon, in which all eight sides have the same length. Inside the regular octagon below, we have drawn three "diagonals"—lines connecting two of the extreme points. How many diagonals are there in all?

Answer, page 240

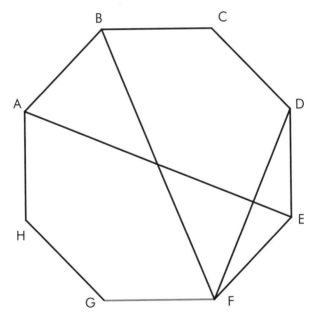

Hint: Be sure not to "double-count" the diagonals. The diagonal joining A to E is the same as the diagonal joining E to A.

On the Trail

One of the numbers below becomes a common English word when converted into Roman numerals. Which one?

38	54	626
1,009	2,376	3,128

Answer, page 240

From Start to Finish

Imagine that the diagram below represents city blocks. The idea is to walk from point S to point F, and of course you can only walk along the lines. The entire trip is five blocks long. In how many different ways can you make the trip?

Answer, page 240

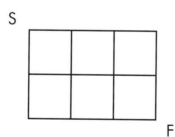

S

F

Hint (On the Trail): Remember that I = 1, V = 5, X = 10, L = 50, C = 100, D = 500, and M = 1,000.

Hint (From Start to Finish): Let A stand for across and D for down. How many different combinations can you make of 3 A's and 2 D's? (For example, AAADD corresponds to traveling from the Start to the Finish along the outermost route. There are many other routes.)

Apple Picking

Seventh Heaven Orchards decides to hold a special sale at the end of the season, hoping that people will come and buy the apples that have already fallen from the trees! They decide on an unusual system for pricing the apples. The bags they give out hold just seven apples each. The orchard then charges its customers five cents for every bag of seven apples, and 15 cents for every apple left over!

According to this system, which costs the most: 10 apples, 30 apples, or 50 apples?

Answer, page 241

Hint: There is nothing tricky about the calculations here, but the answer may be a surprise.

Playing the Triangle

The triangle in the diagram has the lengths of two sides labeled. The reason the third side isn't labeled is that the labeler couldn't remember whether that side was 5 units long, 11 units long, or 21 units long. Can you figure out which it is? (Sorry, but the diagram is not drawn to scale!) *Answer, page 241*

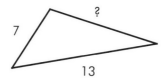

Generation Gap

Grandpa Jones has four grandchildren. Each grandchild is precisely one year older than the next oldest one. One year Jones noticed that if you added the ages of his four grandchildren, you would get his age. How old is Grandpa Jones?

A) 76

B) 78

C) 80

Answer, page 242

Hint (Playing the Triangle): It is not possible to take just any three numbers and form a triangle with those numbers as the lengths of the sides. Remember, the shortest path between two points is a straight line.

Hint (Generation Gap): Trial and error may work out here. You might also save some time if you notice that one of the three ages has a property that the other two do not have.

The Birthday Surprise

A math professor was lecturing his students on a remarkable fact in the world of probabilities. The professor noted that there were 23 students in the class, which meant that the likelihood that some two people in the room shared a birthday was 50 percent!

The professor expected the students to be surprised—most people figure that you'd need many more people before you'd have a 50% chance of a shared birthday. Yet the class wasn't surprised at all. In fact, one student claimed that the professor had miscalculated, and that the likelihood of a shared birthday in the room was in fact much greater than 50%.

What had the professor overlooked?

Answer, page 242

Hint: For this particular class, the likelihood of a shared birthday was almost 100%!

The Run-Off

In a 10-kilometer race, Alex beat Burt by 20 meters and Carl by 40 meters. If Burt and Carl were to run a 10-K race, and Burt were to give Carl a 20-meter head start, who would probably win?

Answer, page 242

Hint: It's a trick question. The trick is to see that the race between Burt and Carl will not be a tie.

The French Connection

Jason and Sandy took five tests during their first year in French class. Jason's scores were 72, 85, 76, 81, and 91. Sandy's scores were 94, 79, 84, 75, and 88. How much higher was Sandy's average score than Jason's average score?

Answer, page 242

Hint: You can figure out the average grade for each student by adding up the individual test scores and dividing by five. But if you look closely at the test scores, you may find a shortcut.

Mirror Time

Below is the digital display of a clock reading four minutes after four. As you can see, the hour and minute figures are the same. It takes one hour and one minute before you see this pattern again—at 5:05.

What is the shortest possible time between two different readings of this same type?

Answer, page 243

$$4{:}04$$

Staying in Shape

The figure below shows one way to join four squares at the edges and make a solid shape. How many different shapes can be created out of four squares? (Two shapes are not considered different if one can simply be rotated to produce the other.)

Answer, page 243

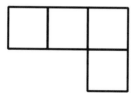

Hint (Staying in Shape): Pencil and paper are really required for this one.

Hint (Mirror Time): There are twelve such times during each twelve-hour span. Going from 1:01 to 2:02 takes one hour and one minute; same for going from 2:02 to 3:03. But there is one occasion when the time gets shorter.

Going Crackers

A cracker company isn't pleased when it finds out the results of a survey it has taken. According to the survey, although customers would rather have a cracker than have nothing at all, those same customers would prefer peanuts to anything else!

A junior employee at the company decides that this is his big chance for a promotion. He claims to his boss that what the survey really said was that customers prefer crackers to peanuts. How in the world could he come to that conclusion?

Answer, page 243

Hint: The secret to this one is in the wording. You have to find something that is both "better" than peanuts and "worse" than crackers. Sometimes there's nothing tougher than a good logic problem!

The Missing Shekel

A farmer in ancient Transylvania took his rutabagas to market each week. His standard price was three rutabagas for a shekel. On an average week, he sold 30 rutabagas and came home with 10 shekels.

One week, he agreed to sell the rutabagas grown by his neighbor, who wasn't feeling well enough to make the trip into town. The only surprise was that the neighbor's preferred price was two rutabagas for a shekel. When the neighbor sold 30 rutabagas, he came home with 15 shekels.

The farmer decided that the only fair thing to do was to sell the combined crop at the rate of five rutabagas for two shekels. But when he added up his money after selling both his crop and his neighbor's crop, he had only 24 shekels, not the 25 he was expecting.

What happened to the missing shekel?

Answer, page 243

Hint: Is the price of five rutabagas for two shekels as fair as it looks?

Quarter Horses

Two horses live on a large piece of land shaped like a quarter-circle. The horses' owner wants to give each horse its separate space by building a fence on the property, but it is important that the two horses have the same space in which to run around. Below are three ways in which a straight fence can be installed to divide the area precisely in two. Which fence is the shortest? Which is the longest? *Answer, page 244*

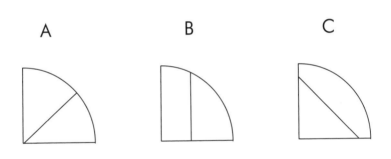

A   B   C

Hint: The simplest thing to do is to compare each length to the radius of the circle. The radius of the circle is the length of either of the straight lines that form the quarter-circle.

MAGICAL MATH

INTRODUCTION

This section uses mathematical principles to present demonstrations that appear to be magic. In many, the math principle is well concealed. In fact, the use of numbers may not even be apparent. As for the tricks, you'll perform extraordinary effects with cards, dice, or pencil and paper. No sleight of hand is used here, but misdirection, clever presentation, and other subtleties ensure that the spectators will be completely flummoxed. You'll read minds, make accurate predictions, or discover a person's age. And you'll show the power of your brain by performing lightning-quick calculations, constructing apparently complex "magic squares," and demonstrating your remarkable memory in a variety of ways.

To increase the excitement when you perform these tricks, avoid treating them as puzzles involving numbers rather than magic. You don't want the spectators dismissing the tricks by saying to themselves, "I don't know exactly how it's done, but I know that it's just mathematics." The best way to keep this from happening is with your running patter. You will find patter ideas are provided for most of these tricks. Well-thought-out patter can set the stage and turn nearly any trick into magic.

Some of the tricks are followed by an explanation called "Why It Works." These explanations will not only help you understand the principle behind the trick, but might cause other ideas to occur to you. You may then want to consider "customizing" the trick by making alterations to it, or even find you are developing other new tricks of your own.

Enjoy!

BAFFLING TRICKS

A Sly Inference

Bob Hummer invented this excellent trick, which, basically, is an exercise in simple logic. Required are three different small objects. In our example, we will assume the objects are a key, a pencil, and a ring.

Lorna is eager to assist, so tell her, "Let's try a test to see if I can discover the one object of three that you are thinking of. We'll use these three objects." Place the three objects in a row on the table. The order doesn't matter. "Lorna, here we have a key, a pencil, and a ring." Let us assume that you have set them on the table so that the key is on your right, the pencil is in the middle, and the ring is on your left. Point to the ring, saying, "Right now the ring is in Position 1." Point to the pencil. "The pencil is in Position 2." Point to the key. "And the key is in Position 3." As you can see, you number the positions so that, from her view, they are in ascending order from left to right; from your view, therefore, they are in descending order from left to right *(Illus. 1)*.

"Lorna, I'll turn my back and then ask you to exchange two of the objects. Now, don't tell me which objects you exchange, just tell me the positions. You might, for instance, change the positions of the ring and the pencil. The ring is in Position 1 and the pencil is in Position 2. Just tell me, 'One and two.'"

Make sure that she understands. Turn your back and tell her to begin. After she exchanges two objects, giving you their original positions, tell her, "Now, again exchange any two of the objects, once more telling me which positions you're switching." She continues doing this for as long as she wishes.

When she is done, tell her, "Lorna, simply think of one of the objects. Are you thinking of one? Good. Please remember that object. Now, without telling me anything, switch the positions of the other two objects. In other words, exchange the other two objects."

After she does so, say, "Let's go back to the original procedure. Exchange two objects and tell me which positions you switched." She continues doing this any number of times. When she says she

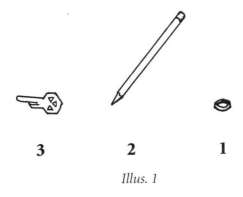

3   **2**   **1**

Illus. 1

is done, turn and face the group. "Lorna, please concentrate on the object you thought of." You carefully study all three objects. Finally, your hand falls on one of them. "This one," you declare. And, of course, you're right.

You repeat the stunt any number of times to prove that this is not mere coincidence, that you have some sort of extraordinary power. There is no point in telling the group that this extraordinary power is the ability to count on your fingers.

The secret: At the beginning, note the middle object on the table, the one at Position 2. In our example, this is the pencil. Turn your back and hold up the first three fingers of either hand. Mentally, number the fingers 1, 2, and 3. Place your thumb on the middle one of the three fingers, thus marking the pencil as being in Position 2 *(Illus. 2)*. The spectator tells you of an exchange, giving you the position numbers. Using your thumb and three fingers, keep track of the object originally at Position 2—in this instance, the pencil. For example, Lorna first tells you that she has exchanged the objects at Positions 2 and 3. You move your thumb to finger number 3.

Illus. 2

She announces that she has exchanged the objects at Positions 1 and 2. You keep your thumb right where it is, because that's where the pencil remains.

She announces that she has exchanged the objects at Positions 1 and 3. Move your thumb to finger number 1, for the pencil now is in Position 1.

You tell Lorna to think of one of the objects and then exchange the other two objects. Then she is to continue exchanging and announcing as in the beginning. You continue to mark the position of the pencil (the original middle object) as though nothing has happened. As you will see, it doesn't matter whether the pencil is actually in that position.

Finally, Lorna is done. You note the number of the finger on which your thumb rests. Turn around and look at the objects. (Remember that Position 1 is on your right, and Position 3 is on your left.) If the pencil is at the same position as the last position marked by your thumb, then the pencil is the selected object. For instance, when Lorna finished, your thumb marked finger number 3; the pencil lies at Position 3. Therefore, the pencil is the chosen object.

But suppose you end up marking finger number 3, and an object other than the pencil is at that position. For example, suppose the key is at Position 3. In this instance, the selected object would be the ring. In the same way, if the ring were at Position 3, the selected object would be the key.

In short: You end up marking finger number 3. If the pencil is at that position, it is the selection. If the pencil is not at that position, it is eliminated. Also eliminated is the object that rests at that position.

Why It Works: You tell Lorna to think of an object and exchange the other two. Suppose that at this point you're marking the pencil as at Position 3. If Lorna decides to think of the pencil, she'll exchange the other two objects—the ring and the key. Lorna once more begins exchanging objects and notifying you of the moves. As she does so, you'll actually continue to mark the position of the pencil. So when she stops, the pencil will be at the position you're marking on your fingers. She must, therefore, have selected the pencil.

Suppose, however, she thinks of the key. The pencil and the ring are exchanged. So in succeeding moves you're marking the position of the ring. At the end, when you see that the ring is at the number you've been marking, you know that the chosen object is not the pencil. Nor can it be the ring, for the ring was exchanged with the pencil. So it must be the key.

In the same way, if the key is at the position you've been marking, you deduce that the choice can't be the pencil or the key; it must be the ring.

The Easy Way

This is essentially an easier version of the previous trick. It can be performed just about anywhere—at your home, at a friend's home, or even in a restaurant. All that is required is three cups and a volunteer who is likely to have a dollar. Linda has a dollar, so get her to help you perform this trick.

Turn the cups upside down and set them side by side. As you're doing this, note the surface of the cups carefully; one of them will probably have a distinguishing mark of some sort. It could be a dot, a slight discoloration, an uneven ridge, whatever *(Illus. 3)*. Note which cup has this mark. Let's suppose that this mark is on the cup to your left; think of this cup as being in Position 1. The cup to the right of this is in Position 2. And the cup to the far right is in Position 3. (If the marked cup were in the center position, you'd think of it as in Position 2; if it were on the right, you'd think of it as in Position 3. This positioning is different from that used in the previous trick where, from your viewpoint, the numbers were reversed so that the spectator could read the numbers from left to right.)

Distinquishing mark

Illus. 3

Say to Linda, "Here we have three cups in a row. In a moment I'll turn my back. When I do, I'd like you to take a dollar bill, crumple it up, and put it under one of the cups. Then exchange the positions of the other two cups. For example, if you put the dollar bill under this cup..." point to the middle cup, "...then you would exchange the two outside cups." Indicate with your hands how the exchange would be made, but keep the cups in their original order.

Turn away until Linda finishes her task. "The question is, which cup has the dollar under it? I'll choose this one." You turn over the correct cup.

How do you do it? Pretty much the same way as in the previous trick.

All you need do here is figure out which cup has not been moved. Clearly, this will be the correct cup.

In our example, the marked cup is at Position 1. If it's still at that position, then it must have the dollar bill beneath it.

If the marked cup is at Position 2, it means that the cups at Position 1 and Position 2 were exchanged. Therefore, the cup at Position 3 has not been moved and must have the dollar beneath it.

In the same way, if the marked cup is at Position 3, then the cups at Position 1 and Position 3 have been exchanged. The cup at Position 2 must have remained stationary and must, therefore, be the correct one.

Why It Works: Here is a slightly different way of putting the explanation: When you turn back, you look for the marked cup. If the marked cup is still in the same position (Position 1), it is obviously the correct cup.

Suppose, however, that the marked cup is in Position 2. Then it must have been exchanged with the cup that was in Position 2. This means that the correct cup must be in Position 3.

In the same way, if the marked cup is in Position 3, then it must have been exchanged with the cup at that position. Therefore, the cup at Position 2 has not been moved, so it must be the correct one.

Hummer by Phone

Magicians for years have labored over the Bob Hummer principle used in the two previous tricks. Sam Schwartz came up with a variation which is in many ways the best of the lot. In this version, no object is marked. And the method is puzzling even to those who know the original trick. What's more, the trick may be done over the phone! For clarity and increased interest, I have added a patter theme using three face cards.

All that's required is a deck of cards and a coin (or other small object). So give Marty a phone call, and tell him, with suitable pauses, "Take from the deck a face-card family—a king for the father, a queen for the mother, and a jack for the son. It doesn't matter what the suits are. Place the jack faceup on the table to your left. The jack is in Position 1. Place the queen faceup to the right of the jack; she's in Position 2. And put the king faceup to the right of the queen; he's in Position 3.

"You're going to give some money to one of the members of the family—the son, the mother, or the father. So just take a coin and place it on top of one of the three cards, whichever you think is the most deserving.

"Exchange the other two cards. For instance, if you placed the coin on the queen, exchange the jack and king."

You do not ask Marty for any information about this initial move.

"Now, exchange any two cards. Just tell me what positions you're exchanging. For instance, if you're switching the card now at Position 1 with the card at Position 3, just tell me 1 and 3."

Marty does this any number of times.

"Now, perform whatever switches you need to bring the family back together in the proper order—jack at Position 1, queen at Position 2, and king at Position 3. As you do this, be sure to tell me what positions you're switching."

Marty finishes. Without asking a question, you tell him on which card the coin rests. You can repeat the stunt any number of times.

It seems absolutely impossible. Yet, your job is quite simple. You start by assuming that the coin is on the card at Position 1. As in "A Sly Inference," you then keep track of this card on your fingers. You hold up the first three fingers of either hand. Mentally number them 1, 2, and 3. To start, place your thumb on the finger designated as 1; this means that, for your purposes, the coin is now at Position 1. Marty tells you of a switch, giving you the position numbers. If the switch involves the card at Position 1, mark the new position on your fingers. Let's say that Marty announces 1 and 3. You move your thumb to the finger designated as 3. The next time Marty switches with the card at 3, you move your thumb to the appropriate finger. (For further explanation, you might read through "A Sly Inference.")

Ultimately, Marty returns the cards to their original positions. And you, of course, continue keeping track of Position 1. Marty announces that he's done. You note which position your thumb is marking. If the thumb is marking Position 1, the card with the coin on it (the jack) is at Position 1. If the thumb is marking the finger designated as 2, the coin lies on the king at Position 3. And, if the thumb rests on the finger designated as 3, the coin lies on the queen at Position 2.

This is important: When your thumb marks finger number 1, the coin is at Position 1. But the other two positions are switched! If your thumb marks number 2, the coin is at Position 3. And if your thumb marks number 3, the coin is at Position 2.

Conclude with some appropriate remark, depending on who received the coin. You might josh Marty about choosing the woman, the queen. Or, if he chooses the jack, you might say, "I suppose you chose the son because he's most in need of money." Or, if he picks the king, you could say, "What makes you think the father needs any more money than he has?"

Why It Works: The basic principle is explained in the "Why It Works" section at the end of "A Sly Inference." Here's the way it works in this instance: Marty places the coin on one of the cards, and then exchanges the other two. You assume that the coin is on the jack, the card at Position 1. You then follow the moves of the card at Position 1. If that card ends up back at Position 1, then the coin must be on it.

Suppose, however, that Marty placed the coin on the queen, the card at Position 2. The other two cards are switched. Therefore, the card originally at Position 3 is now at Position 1, and the card originally at Position 1 is at Position 3. So it is the card that started at Position 3 that you will actually be keeping track of. No matter how many switches are made, eventually the cards are switched back to their original positions. And you'll find that your thumb is marking Position 3. When your thumb marks Position 3, the coin rests on the card at Position 2.

In the same way, if Marty puts the coin on the card at Position

3, the king, the cards at Positions 1 and 2 are exchanged. This puts the card originally at Position 2 at Position 1. And it's the card originally at Position 2 that you'll be keeping track of with your thumb. When Marty says that he's done and that the cards are back in their original order, you'll be marking Position 2 on your fingers. When your thumb marks Position 2, the coin rests on the card at Position 3.

Note: Obviously, cards need not be used in this version; in a pinch, any objects will do.

Getting Along

Here we have a mathematical stunt that purports to be a compatibility test. It's based on an extremely clever card trick.

You'll need the help of a man and a woman, preferably two who are married or friendly.

Let's say that you elicit the aid of Harold and Jan. Explain to them, "I'd like to find out whether you two are really compatible. We'll use numbers in this experiment."

Give each a pad or sheet of paper and a pencil.

"I'd like each of you to jot down a digit—any number from 1 to 9. Now, make sure that neither one of you can see the other person's number."

You, however, have no such restriction. In fact, you make it a point to hand Harold his writing material last. And you hang around long enough to get a glimpse of the digit he jots down.

Meanwhile, you're scrupulously careful to keep your head averted so that you can't possibly see Jan's digit. Stroll some distance away, and then have Jan perform the following:

1. Double your number.
2. Add 2.
3. Multiply by 5.

Next, you have Jan subtract a number. The number is actually Harold's number, subtracted from 10. Let's say that Harold chose the number 8. Subtract it from 10, and you get 2. So the next step is:

4. Subtract 2.

Suppose Jan has chosen the number 3 and, as I said, Harold chose 8. Following your instructions, Jan doubles 3, getting 6. She adds 2, getting 8. She multiplies by 5, getting 40. (Harold's number is 8; you've subtracted 8 from 10, getting 2. You tell Jan to subtract 2 from her number.) She subtracts 2, getting 38.

Say to Jan, "You have two digits in your answer, don't you?" She says yes. "What are the digits?" "Three and eight." "And what's the digit you thought of originally?" "Three."

"Three! What a coincidence! One of the digits you came up with is three, and three was your original digit. And, Harold, what's your digit?" "Eight," Harold replies."

"Eight! That was your other digit, Jan. You also came up with Harold's digit! You two are really compatible."

Why It Works: The first three instructions given to Jan automatically produce a two-digit number, the second of which is zero. The first digit is one more than the digit she first chose. Thus, if she chose 1, the number she comes up with is 20. If she chose 2, the number she comes up with is 30. And so on. At this point in the example above, Jan has come up with 40.

You have subtracted Harold's number, 8, from 10. So, in Step 4, you tell Jan to subtract 2 from her total. As you can see, whatever digit is subtracted will produce Jan's original choice as the first digit. And, happily, the subtraction will also produce Harold's choice as the second digit.

FUN STUFF

It All Adds Down

Write this column of figures and ask someone to add them up one line at a time:

$$1\,0\,0\,0$$
$$2\,0$$
$$1\,0\,3\,0$$
$$1\,0\,0\,0$$
$$1\,0\,3\,0$$
$$2\,0$$

Before we go any further, why don't you give it a try.

Done? Good. What answer did you get? 5000? Good! You just proved that even extraordinarily bright people will get this wrong.

The correct answer is 4100. Don't feel bad; the vast majority get it wrong.

For best effect, jot down the column of numbers on the back of a calling card. When you show the stunt to someone, use another calling card to reveal the top number first, then the second number, and so on (*Illus. 4*).

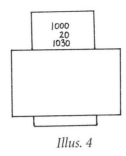

Illus. 4

Why It Works: As if you didn't know! You went like this:

1000	"One thousand...
20	...one thousand twenty...
1030	...two thousand fifty...
1000	...three thousand fifty...
1030	...four thousand eighty...
20	...five thousand."

After progressing through one, two, three, and four thousand, the tendency to go to five thousand is almost irresistible. Of course you know that 80+20=100, but the misdirection is just too strong.

All Together Now

This trick can be wonderful fun for a group. The basic idea is quite old, but Ed Hesse added some deceptive touches.

In preparation, you must write down on a card the number which is double the present year. For instance, a person who had performed the trick in 1996 would have written down 3,992 (1996×2). (It's a small point, but make sure you include the comma after the first digit so that people will be less likely to suspect the actual derivation of the number.)

In performance, make sure everyone in the group has a writing instrument and paper. Then provide these instructions:

1. Please write down the year you were born.
2. Below this, jot down the year of a memorable event—your marriage, your graduation, your discharge from service, whatever.
3. Below this, write down the age you are or will be on your birthday this year.
4. Finally, write down the number of years since that memorable event at the end of this year.
5. Add up all your numbers.

When everyone is done, hold up the card on which you've written your number so that everyone can see it. "How many of you have this number?"

Just about everyone will have it. Only those who are poor at math will miss.

Why It Works: If you take the date on which an event occurred and add to it the number of years ago on which it occurred, you'll end up with this year's date. If you do this twice, you'll end up with a number which is twice this year's date. Which is exactly what happens here.

A Nickel for Your Thoughts

Hand Gary a penny and a nickel. "Gary, I'm going to turn my back. When I do, I'd like you to hold the penny in one hand and the nickel in the other hand."

Turn away. "Gary, please multiply the value of the coin in your left hand by 14." Pause a moment. "Ready?" If he tells you no, wait until he indicates he's ready to continue. If he says yes, proceed immediately.

"Now, multiply the value of the coin in your right hand by 14." He'll tell you when he's done.

"Please add the two numbers together and tell me the total." Unless Gary is seriously deficient in his addition skills, you'll always hear the total 84. You promptly tell him which hand holds what coin.

Since the total is always the same, how do you know this?

The answer is easy. Multiplying 14 by 1 is much easier than multiplying 14 by 5. You have Gary multiply the value of the coin in his left hand by 14. Pause briefly, and then say, "Ready?" If the answer is no, he holds the nickel in that hand. If the answer is yes, the penny is in his left hand.

The Sneaky Serpent

This clever trick is the invention of Karl Fulves.

Three objects are placed in a row on a table. A spectator mixes them. The magician gives the exact position of each object. A beautiful trick, with just one flaw: It works only 5 out of 6 times.

I have adapted the trick to playing cards and have added a devilish patter theme. This version works 6 out of 6 times.

If you wish, you may perform the trick over the phone. But let's suppose you're performing it for a group. Gilbert is a good sport, so ask him to help out.

"Gilbert, I'm going to turn my back in a moment." Hand him the deck of cards. "After I do, I'll provide you with some instructions. If all goes well, I may be able to perform a feat of mind reading."

Turn your back and provide these instructions, pausing at appropriate spots:

"Please take from the deck the K and Q of any suit. These will stand for Adam and Eve. They are, of course, in the Garden of Eden. What's missing? Why, the snake! So please take out the AS (Ace of Spades), who will be Satan, the sneaky serpent. Set the rest of the deck aside.

"Please mix the three cards. Now turn them faceup and deal them into a row. You don't have them in A K Q order, do you?" If the answer is no, say, "I knew that." Then continue your instructions. If the answer is yes, say, "Please don't use that order—it's just too easy." Actually, when the cards are in A K Q order, the trick won't work.

Continue: "Start by switching the serpent with whoever is on his right. If the serpent is on the right end, just leave him be.

"Next, switch Eve with whoever is on her left. If she is at the left end, just leave her be.

"Finally, switch Adam for whoever is on his right. If he's on the right end, just leave him be."

When Gilbert finishes, say, "Let's see if we can arrange to keep Adam and Eve in the Garden of Eden. We'll have to get rid of the serpent. I wonder where he is. I know! It's obvious, isn't it? The serpent most certainly wants to come between Adam and Eve, so he must be in the middle. Please remove him from the middle so he'll stop bothering the happy couple. So there they are, side by side, just as though they're standing at the altar about to get married—Eve on the left, and Adam on the right."

Review:

1. Gilbert removes from the deck the K and Q of any suit, along with the AS.
2. He deals them in a row in any order. But you eliminate the A K Q order.
3. Gilbert makes three switches in this order:
   The A for the card to its right.
   The Q for the card to its left.
   The K for the card to its right.
4. The three cards are now in this order: Q A K.

Why It Works: When you begin, there are 6 possible positions:

   1) A K Q
   2) A Q K
   3) Q A K
   4) Q K A
   5) K A Q
   6) K Q A

You verbally eliminate 1) because you know that, with this setup, the series of moves will not bring about the desired result.

You provide the first instruction: Exchange the A with the card on its right. After Gilbert does so, here are the only possible setups:

1) Q A K
2) Q K A
3) K Q A

The second instruction: Exchange the Q with the card on her left. Now there are only two possibilities:

1) Q A K
2) Q K A

The final instruction: Exchange the K with card on his right. Only one possibility remains:

Q A K

As you can see, in just three clever moves, you've eliminated all possibilities except the desired one.

Notes:

1) The trick may be enhanced if you hesitate and stammer a bit as you provide the directions, creating the impression that you're simply making up each move as you go along. "Let's see. Let's try Eve—the queen. How about exchanging her—oh, I don't know...Maybe...yeah...How about exchanging her for the card on her left."

2) To make sure the trick works, you more or less tell Gilbert not to use the A K Q order. There's at least one other way you can eliminate the A K Q order.

Before providing instructions for the three switches, say, "When you put your three cards in a row, make sure that the serpent is to the right of Adam or Eve—either one. You see, Satan likes to tempt by whispering in the left ear." Then proceed with the instructions, as above.

LIGHTNING CALCULATION

The Speedy Adder

Does the name Leonardo Fibonacci strike a responsive chord in you? Possibly not. It is time for a brief history lesson. This late 12th- and early-13th-century Italian mathematician made amazing discoveries in his field. He is best known, however, for a number sequence known as the Fibonacci series, in which each number is the total of the two previous numbers.

For instance, a number is written down. Another number is written beneath it. The two are added together, and the total placed beneath the second number. Then the second number and the third number are added together and this total is put down below the third number. The sequence can go on indefinitely.

Of what use is this? Using this series, you can perform an astonishing lightning calculation—or at least appear to do so.

If you like, you can introduce the stunt by briefly discussing Fibonacci. Or you might just explain how to develop a Fibonacci series without naming it. Usually, I prefer the latter.

"We're going to develop a rather large number totally by

chance. I'll show you how."

Ask someone to name a small number. Jot it down. Suppose the number is 8.

8

Ask someone else to name a different small number. Let's say the number is 13.

8
13

"Eight and 13 is 21," you point out. "So that would become the next number." You now have this:

8
13
21

"How do we get the next number? We just add the last two together. In this instance, we have 13 and 21. We add them together, and we get 34." You now have this on your paper:

8
13
21
34

To make sure everyone understands, it's time for a brief quiz. "So what would the next number be?" Sure enough, several have worked out that you add 21 and 34 together, getting 55.

Now you're ready to get down to business. Toss your worksheet away. On another sheet, put a column of numbers from 1 to 10 with a dash after each figure *(Illus. 5)*. This has a dual purpose. It makes sure that your assistants put down exactly 10 numbers, and—as you'll see—because you need to be able to spot the 7th number at a glance.

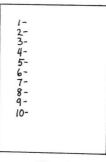

1 –
2 –
3 –
4 –
5 –
6 –
7 –
8 –
9 –
10 –

Illus. 5

Ask Rudy and Julie to assist you. "In a moment, I'll turn my back. After I do, I'd like you each to think of a number. Rudy, think of a number from 5 to 15. Julie, you think of one from 10 to 20. Rudy, please put your number after number 1 on the sheet. Julie, you put yours after number 2 on the sheet. Then we'll have Julie do the hard work. She'll add the two numbers and put the total after number 3 on the sheet. Then she'll continue, all the way down to 10. Rudy, you can be the official referee. Make sure Julie doesn't accidentally put down a wrong number. After you're done, I'll try to add up the column of numbers as quickly as I can."

Turn away while the two do their math exercise. When they're done, turn back. Take the writing instrument and draw a line under the column of figures. After a quick glance down the column, jot down the total. Just as with regular addition, you put in the digits, moving from right to left.

Ask Rudy to add the column and put his answer below yours. Sure enough, his answer is identical to yours.

How do you do it? Nothing to it. Just multiply the 7th number by 11. Please! No crying about, "I'm no good at multiplying in my head." Of course you're not. Neither am I. That's why I worked out an easy way for me to multiply by 11 without hurting my head bone. Let's take a look at a typical Fibonacci series (*Illus. 6*).

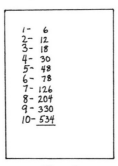

Illus. 6

The correct total is 1386. You can arrive at this total quickly by multiplying the 7th number (126) by 11. First, understand that, with the limitations placed on the choice of numbers, you'll always end up with a four-digit number. Furthermore, the 7th number will always be a three-digit number.

So you're working with the 7th number, 126. If you were to simply multiply by 11, it would look like this:

$$
\begin{array}{r}
1\,2\,6 \\
\underline{\times 1\,1} \\
1\,2\,6 \\
\underline{1\,2\,6} \\
1\,3\,8\,6
\end{array}
$$

As you can see, the first digit on the right will always be the same in both numbers. In this instance, the digit is 6. So you put this down on the right, below the column of figures.

6

To get the digit to the left of this, you add the second and third digits together. The number is 126; we add the 6 and 2, getting 8. So we put down 8 to the left of the 6.

86

How do we get the first two digits? Consider the first two digits of the 7th number. The 7th number is 126, and the first two digits form the number 12. We add to this the first digit, 1. 12+1=13.

1386

Let's try another example *(Illus. 7)*.

The correct answer is 2497. The 7th number is 227. So the number on the far right will be 7.

Add the last two digits of 227. 2+7=9. So 9 is the next digit to the left: 97.

Take the first two digits of 227, and you get the number 22. Add the first digit, which is 2, and you get 24: 2497.

```
1-   15
2-   19
3-   34
4-   53
5-   87
6-  140
7-  227
8-  367
9-  594
10- 961
```

Illus. 7

Let's try one more example to illustrate an exception *(Illus. 8)*.

The correct answer is 1848. The 7th number is 168, so the digit on the far right will be an 8.

The last two digits of 168 are 6 and 8. We add them and get 14. In the previous examples, we had one digit; here we have a two-digit number. No problem, however. You treat it exactly like any

```
1-    8
2-   16
3-   24
4-   40
5-   64
6-  104
7-  168
8-  272
9-  440
10- 712
```

Illus. 8

other addition; that is, you enter the 4, and you carry the 1: 48.

The first two digits of 168 form the number 16. To this, you add the 1 that you're carrying. $16+1=17$. Now, you add to this the first digit, which is also a 1. $17+1=18$: 1848.

Summary:

Two spectators construct a Fibonacci series of 10 numbers. You pretend to add the numbers, but actually multiply the 7th number in the series by 11. This is easy, because you don't have to remember any numbers; you simply look at the 7th number and work out the answer bit by bit.

1. The 7th number will contain three digits.
2. Put down the last digit as the last digit of your answer. If the 7th number is 125, you put down 5 as the digit on the far right.
3. As with regular addition, you place the next digit to the left of the first digit. You get this digit by adding together the 2nd and 3rd digits of the 7th number. Still assuming that the 7th number is 125, we add together the 2 and 5, getting 7. Put 7 to the left of the 5: 75.
4. The first two digits of 125 form the number 12. Add to this the first digit, in this instance 1. $12+1=13$. So put down 13 to the left of the other two digits: 1375.

The only exception:

1. Assume that the 7th number is 194. Put down the 4 on the right.
2. Add the last two digits, getting 13. Put down the 3 and carry the 1: 34.

3. The first two digits form the number 19. Add in the number you carried. $19+1=20$. Now, as before, add the first digit to your total. $20+1=21$. Put this down to the left: 2134.

Why It Works: Obviously, each number in a Fibonacci series is a fraction of the total. It happens that the 7th number is always precisely 1/11th of the final total. No other number in the series provides a consistent result.

Who figured this out, and how? Ahhh...I have a better idea. Take a card, look at it, remember it, put it back.

That's right. I don't know!

An Additional Trick

This is an excellent follow-up to the previous trick. The effect is similar, but the method is completely different. I have changed the original trick a trifle to make it more deceptive.

Say to Roger, "Let's take turns putting down numbers. Then we'll see how fast I can add them up. Let's start by putting down a 5-digit number."

He does so. You write a number below it. But your number is quite special *(Illus. 9)*.

As you put down your number from left to right, you make sure that each digit adds to 9 with the digits just above it. For instance, Roger's first digit is 5. You subtract 5 from 9, getting 4. So your first digit is 4.

You do something different for the last digit on the right. You

Illus. 9

make sure that your digit adds to 10 with the digit above it. Roger
has a 6 as his last digit. You subtract this from 10, getting 4. So you
put down 4 as your last digit.

Have Roger jot down another 5-digit number below the first
two. He does so, and the sheet might look like that in *Illus. 10*.

You place a number below this, again making sure that each of
the first four digits totals 9 with the digit above it, and that the last
digit totals 10 with the digit above it *(Illus. 11)*.

Illus. 10

54 296
45704
93478
6522

Illus. 11

Since Roger's first number is a 9, you put nothing below the digit. Just casually say, "I think I'll try a four-digit number." If one of Roger's interior numbers were a 9, you would place a zero below it.

Again Roger jots down a 5-digit number, and you place a number below it *(Illus. 12)*.

Finally, you say to Roger, "Why don't you put down the last two numbers yourself. To make it harder, put a 5-digit number on top of the column and put another 5-digit number at the bottom of the column. I'll turn away."

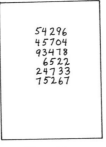

54296
45704
93478
6522
24733
75267

Illus. 12

When Roger's done, turn back, and draw a line below the eight numbers. Then, just about as fast as you can write, jot down the correct total.

However did you manage that? You simply totaled the top and bottom numbers. When you were done, you placed a 3 at the left of your total.

Let's take a look at what Roger left you *(Illus. 13)*.

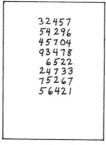

Illus. 13

The top number, which he just added, is 32457. The bottom number, which he also just added, is 56421. You add these two together, putting down your answer as you go.

$$
\begin{array}{r}
32457 \\
+56421 \\
\hline
88878
\end{array}
$$

You have placed one digit in each column, but you'll require 6 digits. So you place a 3 at the front of your total: 388878.

When Roger adds the columns to check your result, he finds that you're exactly right.

You'll always place a 3 at the front of your total, with one exception: Sometimes when you add together the last two digits, those at the extreme left, the result will be a two-digit number. For instance, the result might be 16. When this happens, you enter the second digit in the normal fashion. In this example, you'd jot down the 6. Then you add 3 to the first digit. Since the first digit is always 1, it means that the digit that you place at the front will be 4. Thus:

$$
\begin{array}{r}
7\,3\,4\,8\,1 \text{—Top number} \\
+\,9\,5\,1\,3\,2 \text{—Bottom number} \\
\hline
4\,6\,8\,6\,1\,3
\end{array}
$$

Let's try another example. The spectator has jotted down three 5-digit numbers; you have added a number below each one *(Illus. 14)*.

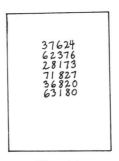

Illus. 14

Note the last two numbers in the column. The spectator wrote 36820. Immediately you notice that the spectator has placed a zero at the end of his number. You're supposed to place a digit there which will add to 10 with the digit above it. Obviously, the only digit that will do is another zero. But when you do this, you have to make sure that your next digit to the left will add to 10 with the digit above it. So you simply start at the left and proceed through the first three digits in the regular way. You make sure that each of your digits adds up to 9 with the digit above it. When you reach the fourth digit from the left (the 2, in this instance), you make sure that your digit adds up to 10 with the digit above it. Finally, put a zero at the end.

The spectator adds two more 5-digit numbers, one above the others and one below the others (*Illus. 15*).

81014
37624
62376
28173
71827
36820
63180
98341

Illus. 15

You draw a line under the column. Then you add together the number at the top of the column and the number at the bottom of the column. Going from right to left, you perform the first four additions:

$$\begin{array}{r} 81014 \\ +98341 \\ \hline 9355 \end{array}$$

The two digits to be added at the far left are 8 and 9. Add them together, getting 17. When you have a two-digit number, enter the second digit—in this instance, 7. Then place a 4 in front of the entire number:

$$
\begin{array}{r}
81014 \\
+98341 \\
\hline
479355
\end{array}
$$

Why It Works: The spectator writes down a 5-digit number; you write a number beneath it. You make sure that each of your first four digits adds to 9 with the digit above it, and that the last digit adds to 10 with the digit above it:

$$
\begin{array}{r}
37862 \\
62138
\end{array}
$$

Because of what you have done, these two numbers add up to 100,000.

You go through the procedure twice more, each time guaranteeing that the total of the two numbers will add up to 100,000.

At this point, you have six numbers entered, and the total of these numbers is 300,000. You turn away, and your assistant puts down two more numbers.

When you turn back, here's what you actually do: You add together the last two numbers that the spectator put down, and you add to that 300,000.

Obviously, you add the 300,000 by placing a 3 in front of the total, or, if the last addition amounts to 10 or more, by placing a 4 in front of the total.

Note: When the spectator puts in his last two numbers, you have him place one on top of the column, and one on the bottom. Why not have him put both at the bottom? If they're both at the bottom, a spectator might more readily see that you're adding the two together.

MEMORY TRICKS

As I Recall

I wish I knew whom to credit for this clever pseudo-memory trick. Regardless, I've added a few refinements.

You'll need the assistance of Mary Lee, who is an excellent card shuffler. Hand her the deck, saying, "Please give these a good shuffle, Mary Lee, because I'd like to demonstrate my ability to memorize numbers."

When she finishes, take the deck back and set it facedown on the table. Hand Mary Lee a pencil and paper. "I'm not good at face cards [a king, queen, or jack], so we'll eliminate those. And I'm still working on suits [spades, hearts, diamonds, and clubs], so we'd better just confine ourselves to the values. Please jot these down in order."

Pick up the deck. Look at the top card, without letting anyone else see it. Give Mary Lee a number to write down; then place the card on the bottom of the deck. Continue doing this until you've provided some 20 numbers.

"Let's see if I can recall those numbers." Give the deck a quick shuffle, mumbling, "We'll make sure there's no chance I'll use the deck."

You then proceed to recite the numbers perfectly.

How can you possibly remember those numbers? You don't. You remember things like your social security number, an old phone number, a birth date.

Let's say your social security number is this: 372-06-9871. Obviously, you'll be able to remember this perfectly. So you simply recite this as you look at each card.

You look at the first card and say, "Three." You place the card on the bottom. You look at the next card and say, "Seven," and place that card on the bottom. You continue through the rest of your social security number. When you come to the zero, you convert it to 10. When you come to the 1, you convert it to ace.

So here are the first nine values you call out to Mary Lee: 3, 7, 2, 10, 6, 9, 8, 7, ace.

Let's say that a familiar phone number is 280-7156. You continue by calling out these values: 2, 8, 10, 7, ace, 5, 6.

Suppose there's a birth date which you know quite well—9/24/79, for instance. You recite these values: 9, 2, 4, 7, 9.

There you have 21 values which apparently you memorized just by glancing through the cards. What a genius! What a mind! What a hoax!

But don't get overconfident. You got away with it—good. But it's not advisable to repeat the trick. People might start to wonder why you're not showing them the cards.

Notes:

1) You've told the group that you'll skip face cards. As you come across a face card, toss it out faceup with a comment like, "No, face cards are too hard."

2) Quite often you'll name a number and, by coincidence, the card you're holding happens to be of that value. Let's say that you turn up the 2H just as you're about to announce the first digit of your phone number, which is also a 2. "Jot down two, please, Mary Lee." Show everyone the 2H, saying, "I won't forget the two of hearts; that's one of my lucky cards." Turn the card facedown and place it on the bottom.

If you're lucky enough to have this occur once or twice while calling out the numbers, the effect is considerably enhanced.

3) As I mentioned, you should use easily recalled numbers, like your social security number, a phone number, a birth date. You can also use a year, an address, or—if you were in service—your service number.

Ah, Yes, I Remember It Well

You present a chart on which you have typed anywhere from 20 to 40 lines of digits. My chart consists of 20 lines, double-spaced. Each line contains 20 digits. Here it is:

```
 (1) 4 5 9 4 3 7 0 7 7 4 1 5 6 1 7 8 5 3 8 1
 (2) 7 9 6 5 1 6 7 3 0 3 3 6 9 5 4 9 3 2 5 7
 (3) 6 9 5 4 9 3 2 5 7 2 9 1 0 1 1 2 3 5 8 3
 (4) 9 3 2 5 7 2 9 1 0 1 1 2 3 5 8 3 1 4 5 9
 (5) 8 3 1 4 5 9 4 3 7 0 7 7 4 1 5 6 1 7 8 5
 (6) 1 7 8 5 3 8 1 9 0 9 9 8 7 5 2 7 9 6 5 1
 (7) 0 7 7 4 1 5 6 1 7 8 5 3 8 1 9 0 9 9 8 7
 (8) 3 1 4 5 9 4 3 7 0 7 7 4 1 5 6 1 7 8 5 3
 (9) 2 1 3 4 7 1 8 9 7 6 3 9 2 1 3 4 7 1 8 9
(10) 5 6 1 7 8 5 3 8 1 9 0 9 9 8 7 5 2 7 9 6
(11) 4 6 0 6 6 2 8 0 8 8 6 4 0 4 4 8 2 0 2 2
(12) 7 0 7 7 4 1 5 6 1 7 8 5 3 8 1 9 0 9 9 8
(13) 6 0 6 6 2 8 0 8 8 6 4 0 4 4 8 2 0 2 2 4
(14) 9 4 3 7 0 7 7 4 1 5 6 1 7 8 5 3 8 1 9 0
(15) 8 4 2 6 8 4 2 6 8 4 2 6 8 4 2 6 8 4 2 6
(16) 1 8 9 7 6 3 9 2 1 3 4 7 1 8 9 7 6 3 9 2
(17) 0 8 8 6 4 0 4 4 8 2 0 2 2 4 6 0 6 6 2 8
(18) 3 2 5 7 2 9 1 0 1 1 2 3 5 8 3 1 4 5 9 4
(19) 2 2 4 6 0 6 6 2 8 0 8 8 6 4 0 4 4 8 2 0
(20) 5 7 2 9 1 0 1 1 2 3 5 8 3 1 4 5 9 4 3 7
```

Each digit on the chart is generated by totaling the two previous digits. For instance, on Line 16, you see that the first two digits are 1 and 8. Obviously, $1+8=9$. So the next digit is 9: 189.

Now, you total 9 and 8, getting 17. But you use only the second digit, 7:1897.

If you look over the chart, you'll see that all the numbers are generated this way.

Hand the chart to Larry, saying, "I've memorized all these lines, Larry. It took me months and months, but I think it was worth it to be able to demonstrate my superb memory."

You turn your back, saying, "Pick out a line, Larry, and tell me which one it is." He tells you. You immediately tell him all the digits on the line. The trick may be repeated.

How? The number of the line tips off all the remaining digits. First, let's assume that the number of the line is a single digit. If the number is odd, you add 3 to it; if the number is even, you add 5 to it. This gives you the first digit.

If Larry tells you he's looking at Line 5, you note that this is an odd number, so you add 3 to it. Your first digit, then, is 8.

You then add the number of the line to the first digit: $5+8=13$. When you have a two-digit number, you always use the second digit only: 83.

Now, you're off and running. As explained earlier, once you know the first two digits, you can generate all the remaining digits.

Suppose that Larry's selected line is a single-digit even number. He chooses Line 4, for instance. You add 5, giving you 9.

You add 4 and 9, getting 13. Once more you use only the second digit, 3: 93.

Larry might choose a two-digit line. As with a single-digit line, you first add 3 to an odd number and 5 to an even number. For instance, Larry selects Line 13. You add 3 to this, getting 16. You use the 6 as the first digit.

To arrive at the next digit, however, you note the two digits in Larry's selection. In this instance, he chose line 13. You add the two digits together. This gives you 4. To this, you add the value of the first digit, 6. 6+4=10. You take the second digit; in this instance, you have a zero: 60.

Suppose Larry chooses an even two-digit number, like 16. You add 5 to it, getting 21. You use the second digit, 1, as the first digit in the line.

Now, you add the digits in 16, getting 7. This, when added to the first digit, gives you 8: 18.

So, you have named the first two digits, and Larry is dumbfounded. You could go on naming digits forever, but that would give away the show. You'd better make sure you stop at 20. Since your back is to the spectators, keep track on your fingers. I start with my left thumb and move from left to right. After I hit my right thumb for the second time, I stop naming digits.

There's no reason you can't perform the trick again. But make sure you take the sheet back when you're finished. Given enough time, some clever rascal might figure out the code.

Notes:

1) Yes, yes, I can hear some of you whining, "Isn't there an easier way to get those first two digits?" Yes, there is. Larry tells you what line he chose. You concentrate and finally admit defeat: "I just can't seem to think of that first digit. What is it?" He tells you. You can now generate the second digit, as described above. Or, if you really don't want to strain yourself, admit to Larry that you can't get the

second digit either. Now, you should be able to finish the line.

2) Practice doing several lines aloud. You'll find that it's surprisingly easy. The fact that you have just said two digits aloud keeps them fresh in your mind as you add them together to form the next digit.

MAGIC SQUARES

An Easy Square

A typical magic square looks like the one shown in *Illus. 16.*

Notice that each column and each row adds up to 15. Also, each diagonal adds up to 15. First, let me explain how to construct this magic square. It may sound a bit complicated, but I'll show you an extremely easy way to construct it. Then I'll provide you with a spectacular trick which demonstrates your astonishing ability with numbers.

Start by putting the number 1 in the top middle space *(Illus. 17)*. Then count 7 boxes to your next number. You count moving from left to right and then to the next lower box on the left...just

8	1	6
3	5	7
4	9	2

Illus. 16

as though you were reading. When you hit the seventh box, put in the next number, 2 *(Illus. 18)*.

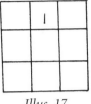

Illus. 17

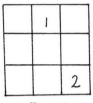

Illus. 18

Starting at the next box, the one at the top left, we count 4 to get to the next box, where we put in the next number, which is 3 *(Illus. 19)*.

Now, from the next box, we count 3 boxes for the placement of the next number, 4 *(Illus. 20)*.

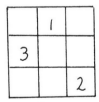

Illus. 19

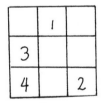

Illus. 20

Then it's back to 7 boxes for the next number, 5 *(Illus. 21)*.
Count 7 boxes again and put in the next number, 6 *(Illus. 22)*.
Count 3 boxes, putting in the next number, 7 *(Illus. 23)*.
Count 4 boxes and put in the next number, 8 *(Illus. 24)*.
Last of all, you would simply fill in the one empty box with the

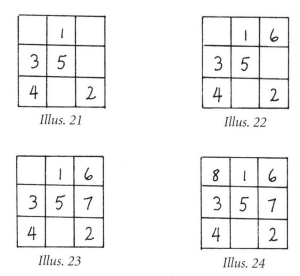

Illus. 21

Illus. 22

Illus. 23

Illus. 24

final number, 9. Actually, it is the 7th box from the preceding number.

To summarize the positioning of the numbers: The first number is placed in the middle box of the top line. After that, we count to succeeding numbers in this order: 7, 4, 3, 7, 7, 3, 4, 7.

The last four counts are a mirror image of the first four counts; this makes memorizing the numbers very easy. It also helps to note that the first number, 7, is followed by 4 and 3—two numbers which add up to 7.

Now you can construct a magic square containing the digits from 1 to 9. You can also now construct a magic square beginning with any number.

How would you present this as a trick? Depending on the size of the group, you can use either a large portable blackboard or a sheet of paper. Let's assume you're using paper and pen. Draw a square with nine empty boxes.

"Although you can't quite see it yet," you say, "this is a magic square. A magic square is one in which the numbers add up to the same total in every possible direction." At this point, have the group choose a representative. Address the nominee: "I'd like you to provide me with any number from 1 to 100. That will be the number I'll start the magic square with."

The spectator names a number, and you place it in the middle box on the top line. You put in succeeding numbers by counting to the appropriate boxes, as indicated. Suppose the number you're given is 82. It is placed as shown in *Illus. 25*.

The next number is placed 7 boxes away (*Illus. 26*).

You end up with the magic square shown in *Illus. 27*.

Illus. 25

Illus. 26

Illus. 27

You then show that, in every direction, the numbers add up to 258.

Notes:

1) You can have the spectators choose any number. The only reason you make it 1 to 100 is to speed up the trick. I have done 1 to 500, but have never quite dared to exhaust an audience's patience by making it 1 to 1000.

2) You're placing the numbers in the boxes right in front of the spectators, but try not to make it obvious that you're counting boxes to arrive at a succeeding number.

On the Square

Stephen Tucker invented a clever and entertaining trick, which makes up the first part of this demonstration. The latter part is a variation of one of Martin Gardner's creations.

On a sheet of paper or a blackboard, you've written this: $(? \times 4) + 34 =$

"Here we have a problem in algebra. Unfortunately, however, we have two unknowns, which would make this extremely difficult to solve."

Hand the sheet, along with a writing instrument, to Jim. After all, he's always bragging about how he got an A in algebra in high school.

"Jim, I'd like you to rewrite the problem and then solve it. First of all, get someone to call out a number to put in for the question mark."

Jim asks someone in the group to contribute a number. Let's say someone yells out 15. Jim puts that in the equation, and then completes the equation so that it looks like this: $(15 \times 4) + 34 =$

Make sure that he writes it down correctly.

"While you're solving that, Jim, I'm going to build a magic square."

On a separate sheet of paper, you have previously placed a blank square made up of 16 squares, with four rows and four columns.

Since 15 was called out, you add 1 to it, making 16. You enter this number in the lowest left square *(Illus. 28)*. Put the next higher number directly above it. Continue this sequence as shown in *Illus. 29*.

19	23	27	31
18	22	26	30
17	21	25	29
16	20	24	28

Illus. 28   *Illus. 29*

Quite often you'll finish your magic square before Jim finishes solving the equation, but it really doesn't matter. When you're both done, say, "What's your answer, Jim?" It's 94: $(15 \times 4) + 34 = 94$.

"There's no doubt, Jim, that if a different number had been put in for that question mark, your answer would have been different, right?" Right.

Shirley is superb at addition, so hand her your diagram, along with a writing instrument.

"Please add up the four corners, Shirley." She does. "What did you get?" She gets 94.

"Add up one of the diagonal rows, please." She gets 94.

"Add up the other diagonal row." She gets 94.

"Add up the four numbers that form a box in the middle." Again 94.

"Undoubtedly, there are many similar combinations," you lie cheerfully. "But let's try something different. Shirley, please circle any one of the numbers." She does so. "Now, cross out the rest of the numbers in that row and in that column." Suppose she circles number 21. Make sure she properly crosses out the other numbers (*Illus. 30*).

Have her circle another number, and then cross out all the other numbers in that row and column which have not yet been crossed out. Suppose she chooses 24 (*Illus. 31*).

19	23	27	31
18	22	26	30
17	(21)	25	29
16	20	24	28

Illus. 30

19	23	21	31
18	22	26	30
17	(21)	25	29
16	20	(24)	28

Illus. 31

Have Shirley circle yet another number and again cross out all the other numbers in that row and column which have not yet been crossed out. Let's say she chooses 30 *(Illus. 32)*.

Now, tell her, "One other number remains uncircled, Shirley. Please put a circle around that last number." After she does, say, "Please add up the four circled numbers." She does so.

"Remember, Shirley, you freely selected the numbers to be circled. Is it possible that you ended up with the magic total?" It is possible. The total of the four circled numbers is 94.

19	2̶3̶	2̶1̶	3̶1̶
1̶4̶	2̶2̶	2̶6̶	⑨30
2̶7̶	㉑	2̶5̶	2̶9̶
1̶6̶	2̶0̶	㉔	2̶8̶

Illus. 32

Why It Works: My guess is that Stephen Tucker worked backward in developing this trick. It makes sense that he might jot down this fairly simple square, and then notice the properties of it—that the corners added to the same number, and so on. But how do you make a trick out of this? One solution is to develop an equation which will yield a number you can work with regardless of what number is chosen by the spectator. Undoubtedly, this involved considerable trial and error.

The second part of the trick is quite ingenious. Let's take a look at another square. Suppose the spectator put in the number 7 for the question mark. You add 1 to this, getting 8. The number in the lowest left corner will be 8, and the other numbers will follow as before *(Illus. 33)*. The four corner numbers add up to 62. And, of course, each diagonal row adds up to 62, and so do the four numbers in the middle.

11	15	19	23
10	14	18	22
9	13	17	21
8	12	16	20

Illus. 33

You're now going to choose a number and eliminate the other numbers in that row and column. So you choose 19 *(Illus. 34)*.

Under the rules, what numbers are left to pick? You could choose 10, 13, and 20. Or 10, 12, and 21. Or 9, 14, and 20. Or 9, 12, and 22. Or 8, 14, and 21. Or 8, 13, and 22. What do you notice

11	15	⑲	23
10	14	18	22
9	13	17	21
8	12	16	20

Illus. 34

about all the possibilities? That's right; they all add up to 43. Together with 19, each one totals 62.

Let's say that your next choice is 13. The square now looks like *Illus. 35.*

Illus. 35

Notice what your remaining choices are: You may choose either 10 and 20 or 22 and 8. In either instance, the numbers add up to 30. You have already chosen 19 and 13, which total 32. Add 30, and you get 62.

In fact, no matter how you slice it, you get 62. When a square is constructed like this, and you choose numbers in this fashion, you'll always end up with the same number.

So Where's the Money?

A well-known magician came up with an unusual trick using a magic square. But it was really not a trick. The spectator was to seem randomly to choose a square, whereas he was actually forced to choose a particular square. The spectator's instructions told him, step by step, to eliminate every square except the desired one. I felt that this could be improved on.

You'll need playing cards and a coin. (An alternative with pencil and paper is presented in the note at the end.) Remove from the deck A 2 3 4 5 6 7 8 9 of any suit.

Ask Lydia to help you. Say to the group, "Let me tell you a story instead of performing a stunt that uses trickery and deceit.

"Once upon a time, Madame Anastasia, that superb psychic, walked into the police station and told the police, 'I've had a vision. Some stolen money is hidden in one of the cabins at the Moldy Motel.' Madame Anastasia had helped the police before, so they went with her to the motel. When they arrived, they discovered that it was a very old motel with nine cabins."

Deal out the nine cards faceup so that, from Lydia's viewpoint, they look like this:

$$A \quad 2 \quad 3$$
$$4 \quad 5 \quad 6$$
$$7 \quad 8 \quad 9$$

"Lydia, I'd like to see if you have psychic powers—if you can do as good a job as Madame Anastasia did. These cards represent the nine cabins. The ace represents cabin Number 1, the 2 is cabin Number 2, and so on. Madame Anastasia strolled among the cabins and eliminated them until only one was left.

"I'd like you to do the same thing, following this guide which was provided to me by Madame Anastasia herself."

Set a sheet of paper on the table so that Lydia can read it.

"On the other side of this sheet is the number of the cabin that Madame Anastasia finally chose. Let's see if you end up with the same one." Hand her the coin. "This coin will represent Madame Anastasia as she wanders around. Start by placing the coin on any one of the cabins." Pause, as she does so. "Then follow the instructions. But first, you have to know what a move is. Wherever the coin is, you can move directly left or right, or you can move straight up or down. But you can't move at an angle."

The trick will not work unless Lydia starts out on an odd number. If the coin is on an odd number at this point, simply continue. If it's not on an odd number, say, "Would you please make a move, Lydia, so I can be sure I've made myself clear." (Don't say anything like, "I want to make sure you understand." This can be insulting.) After the move, she has to be on an odd number.

"When you follow the directions, Lydia, we'll find out the cabin of your choice."

Tell Lydia to follow the directions on the sheet, which should have something like this typed on one side:

MADAME ANASTASIA'S PERSONAL INSTRUCTIONS

Make your move or moves as indicated in the numbered directions below (you must always land on a motel that has not been eliminated, so make sure you have a safe landing before you actually move the coin). Eliminate the cabin on which the coin lands by turning the card facedown. Place the coin back on that card. Next, eliminate cabin number indicated by turning that card facedown.

1. Make 1 move with the coin. Eliminate Cabin 9.
2. Make 2 moves. Eliminate Cabin 5.
3. Make 4 moves. Eliminate Cabin 7.
4. Make 2 moves. Eliminate Cabin 1.

Lydia ends up on Cabin 3. Ask her to turn the sheet of paper over and read the message aloud. She reads: "Congratulations. Madame Anastasia also chose Cabin 3. Unfortunately, no money was there, so she is now reading tea leaves at a small lunch-room in Sydney, Australia."

Why It Works: This is extraordinarily simple. The instructions force the spectator to land on even numbers, which are eliminated. Meanwhile, the directions eliminate all the odd numbers except the number 3.

Note: As I mentioned, the trick can also be done with paper and pencil. Instead of using playing cards, make a 3×3 box with the numbers 1 to 9 enclosed (Illus. 36). Again, the spectator uses a coin, but eliminates cabins by crossing them out with the pencil.

You use the same instruction sheet, but your introductory remarks will vary a little.

Illus. 36

GREAT MATH CHALLENGES

Circular Logic

The diagram below features four houses, labeled C, X, Y, and Z. One of the houses—C—is located at the center of a circle, and the other three—X, Y, and Z—are somewhere on the circumference. There are lines connecting various houses, as shown.

A member of family Y decides to take a walk starting from his house going clockwise around the circular path. Meanwhile, another member of family Y walks along the straight path, visiting houses X, C, and Z in order—but not stopping—before returning home.

Assuming that the two members of family Y walk at the exact same rate, who arrives home first? *Answer, page 246*

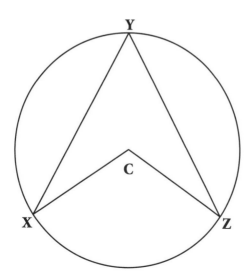

A Roman Knows

Equations of the following sort are called either *cryptarithms* or *alphametics*. Whatever you want to call them, the idea is to substitute a number for each letter so that the indicated arithmetic is correct. No two distinct letters can be given the same number, and once a number has been substituted for a letter, it must substitute for each appearance of that letter. Also, no letter on the left side of any number can represent 0.

The following alphametic isn't particularly difficult, but it is special because the equation is true in Roman numerals; that is, $44 \times 10 = 440$. Your challenge is to make the equation true with regular numbers as well. There are two solutions.

Answer, page 245

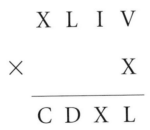

Page Boy

As Jack finished a section of the novel he was reading, his wife posed a curious question: "Dear, suppose you took the page numbers of the section you just read and added them together. What would you get?"

Jack summed them rather hastily and said, "Well, I may have to double-check my arithmetic. It's either 412 or 512, I'm not sure."

Which was it? *Answer, page 245*

No Calculators, Please

Prove the following inequality without multiplying the whole mess out.

$$(1/2) \times (3/4) \times (5/6) \times (7/8) \times ... \times (97/98) \times (99/100) < 1/10$$

Answer, page 246

Born Under a Bad Sign

Sue Perstition was born on Friday the 13th. She recently celebrated one of the birthdays below on Friday the 13th as well. Which one was it? *Answer, page 247*

- A) 10th
- B) 20th
- C) 30th
- D) 40th
- E) 50th
- F) 60th

One, Two, Three

Using all of the digits from 1 to 9 once each, create three 3-digit numbers that are in a ratio of 1:2:3. There are four solutions.
Answer, page 248

Don't Make My Brown Eyes Blue

On a faraway island there was a kingdom in which the inhabitants had either blue eyes or brown eyes. It was a small island, so that with every passing day, everyone on the island saw everyone else. However, there were no mirrors or anything else that reflected, so no one knew the color of his or her own eyes. One day the king decreed the following: 1) At least one of you has blue eyes, and 2) if you wake up in the morning and realize you have blue eyes, you must leave the island at once, without letting anyone see you.

Well, it turns out that there were precisely ten people on the island with blue eyes. Given that everyone on the island was capable of perfect logical reasoning and that everybody on the island knew about the logical abilities of their fellow islanders, what do you suppose happened to them following the king's decree?
Answer, page 248

The Stamp Collection

Phil Atelist has a stamp collection consisting of three books. The first book contains ⅕ of the total number of stamps, the second book contains some number of sevenths of the total number of stamps (he can't remember how many), and the third book contains 303 stamps. How many stamps are in the entire collection? *Answer, page 248*

Too Close to Call

Which is bigger? (No calculators allowed!)
Answer, page 249

$$\sqrt[10]{10} \quad \text{or} \quad \sqrt[3]{2}$$

The Long String

Are there ever 1,000,000 consecutive composite numbers?
Answer, page 249

First Class Letters

Define the "letter class" of a whole number as the number of letters in that number: For example, the letter class of 16 (SIXTEEN) is 7, the letter class of 25 (TWENTY-FIVE) is 10, and so on.

There is only one number between 1 and 5,000 that is the *only* representative of its letter class. Care to find it?

Answer, page 250

Most Valuable Puzzle

When the Sprocketball Writers Association of America votes on the annual Most Valuable Player award, three players are nominated and 10 writers designate a player as first, second, or third place, with each first-place vote worth three points, each second-place vote worth two points, and each third-place vote worth one point. Only the three nominated players can receive votes. (Somewhat different from the procedures followed by the Baseball Writers Association of America, but that's sprocketball for you.)

Under these conditions,

1) What is the smallest number of first-place votes that it would take to clinch the MVP?

2) What is the smallest number of first-place votes a candidate could receive and still win the MVP?

Answer, page 250

Performance Anxiety

Some years ago a leading brokerage firm touted its selections for the just-concluded year by pointing out that its group of "single-best" picks from its top analysts outperformed the S&P 500 by 32 percent. Given that the S&P 500 rose 28 percent for the year in question, that was no mean feat. Only problem was, the firm's selections didn't actually increase by 60 percent, as you might now be thinking. What do you suppose the actual performance was?

Answer, page 251

Double Trouble

Place the numbers one through nine in the boxes below so that *both* multiplications are true.

Answer, page 251

Cereal Serial

A cereal company places prizes in its cereal boxes. There are four different prizes distributed evenly over all the boxes that the company produces. On average, how many boxes of cereal would you need to buy before you collected a complete set?

Answer, page 251

Pairing Off

It is easy to see that the ordered pair (3, 2) satisfies the equation $x^2 - 2y^2 = 1$. But what is the next-smallest pair of positive integers that satisfies the same equation?

Answer, page 251

A Square Deal

A rich man bequeathed a square plot of land to his three children. The number of miles on each side of the plot was not recorded, but it was a whole number. To the first child, the man gave a square plot in the upper-right corner; again, its size was not recorded, but each side was a whole number. To his second child, he gave a plot that was square but for the upper-right corner that the first child owned; again, the sides of this plot were all whole numbers. The third child received the remaining plot, which turned out to be precisely equal in area to the second child's plot. The figure on the next page shows what the plots looked like, although it is not necessarily to scale.

This information is not enough to determine the precise dimensions of each of the plots, but if you know that the first child received the smallest piece possible based on this information, you now have enough to figure out the dimensions of all three of the plots. How big were they?

Answer, page 252

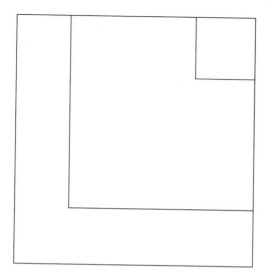

Survival of the Splittest

A population starts with a single amoeba. Suppose there is a ¾ probability that the amoeba will split to create two amoebas, and a ¼ probability that it will be unable to reproduce itself, in which case it will die out. Assuming that all future generations of amoebas have the same probabilities associated with them, what is the probability that the family tree of the single amoeba will go on forever?

Answer, page 253

Shutting the Eye

The figure below may look somewhat familiar. Suppose the radius of the big circle is three times the radius of the small circle. Furthermore, suppose that the "eyelids" are made up of two separate arcs of an even larger circle. What is the length of the radius of that third circle relative to the "pupil" in the diagram?

Answer, page 253

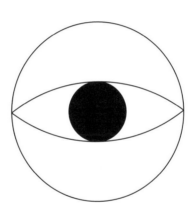

What's in a Name?

As you can see, each of the following three alphametics involves a famous name. Unfortunately, only one of these equations is possible. Can you determine which one has a solution? (For the basic rules of alphametics, see "When in Rome," p. 184.)

Answer, page 254

```
    M A R Y
+   T Y L E R
-----------
    M O O R E


    J O Y C E
+   C A R O L
-----------
    O A T E S


    J A M E S
-     E A R L
-----------
    J O N E S
```

Thumbs Down

It turns out to be impossible to "reverse" a number by multiplying it by 2. In other words, there is no number of the form abcd, for example, such that abcd × 2 = dcba. (The equation is not only impossible for four-digit numbers, it is impossible for *all* numbers.)

However, there is a three-digit number abc *in base 8* such that abc × 2 = cba. Can you find that number?

Answer, page 255

Ticket to Ride

Each of the railroad stations in a certain area sells tickets to every other station on the line. This practice was continued when several new stations were added, and 52 additional sets of tickets had to be printed. How many stations were there originally, and how many new ones had to be added?

Answer, page 255

Going Off on a Tangent

In the diagram below, a circle is inscribed in an isosceles trapezoid whose parallel sides have lengths 8 and 18, as indicated. What is the diameter of the circle?

Answer, page 256

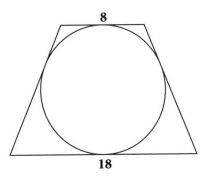

Surprise Ending

For what integral values of n is the expression $1^n + 2^n + 3^n + 4^n$ divisible by 5? *Answer, page 257*

Thanksgiving Feast

On the first Thanksgiving in Plymouth, Massachusetts, legend has it (a just-created legend, that is) that 90 percent of the pilgrims had turkey, 80 percent had corn, 70 percent had pumpkin pie, and 60 percent had mince pie. No one, however, had all four items. Given the limited menu, what percentage of pilgrims had at least one of the two desserts?

Answer, page 257

All About Pythagoras

Recall that a Pythagorean triple is a set of three positive integers that can form the sides of a right triangle. The best known example is the famous 3–4–5 right triangle below:

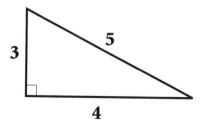

These dimensions create a right triangle because, like any self-respecting Pythagorean triple, they satisfy the Pythagorean theorem: $3^2 + 4^2 = 5^2$

(1) What positive integers cannot be part of a Pythagorean triple?

(2) What is the smallest number that can be used in all three positions—as the hypotenuse, as the longer leg, and as the shorter leg—in three different right triangles?

Answer, page 258

A Decent Decade

Only once in American history has there been a decade in which four of the years were prime numbers. Can you find it?

Answer, page 259

Composite Sketch

If the previous problem was too hard, you might want to tackle this one instead: Can you find the first decade (A.D., of course) in which all of the years were composite? To start you off on the right track, you should know that America was not around during this decade. *Answer, page 260*

Five Squares to Two

Using just two straight cuts, divide the figure below into three pieces and reassemble those pieces to form a rectangle twice as long as it is wide. *Answer, page 260*

Equality, Fraternity

1) Consider the "equation" 6145 − 1 = 6143. Can you move two digits so as to create a valid equality?

2) The equation −127 = −127 is, of course, true. Suppose you were told to move two digits so as to leave a valid equation. You

might try moving both 2's to the end, creating the equation $-172 = -172$, but you've probably guessed that you have to move two digits on the same side of the equation. Any ideas?

Answer, page 260

Blue on Blue

A hat contains a number N of blue balls and red balls. If five balls are removed randomly from the hat, the probability is precisely ½ that all five balls are blue. What is the smallest value of N for which this is possible?

Answer, page 261

Stick Figures

Consider the following equation, which is obviously false.

$$1 1 = 1 1 3 3 5 5$$

A) Can you insert four line segments into this equation to make it correct?

B) Same question, but now you are given only three line segments to work with. To give you a break, the equation doesn't have to be exact, but it does have to be accurate to six decimal places!

Answer, page 261

Heralding Loyd

Sam Loyd was perhaps the greatest puzzle maker in American history. One of his creations was an elegant puzzle that concerned crossing a waterway by boat. You should enjoy its beautiful simplicity.

Two ferryboats traveling at a constant speed start moving at the same instant from opposite sides of the Hudson River, one going from New York City to Jersey City and the other from Jersey City to New York City. They pass one another at a point 720 yards from the New York shore.

After arriving at their respective destinations, each boat spends precisely 10 minutes at the opposite shore to change passengers before switching directions. On the return trip, the two boats meet at a point 400 yards from the Jersey shore.

What is the width of the river?

Answer, page 261

Lots of Confusion

You have another chance to match wits with Sam Loyd. This Loyd original concerns the case of a real estate mogul who bought a piece of land for $243, divided it into equal lots, and sold the entire package for $18 per lot. (If the prices seem low, remember that Loyd lived in the 19th century.) The mogul's profit on the entire deal was equal to what six of the lots had originally cost him.

How many lots were in the piece of land? *Answer, page 262*

Oh, Henry!

Speaking of famous puzzle makers and lots of land, Henry Dudeney (1847–1930) was one of the greatest puzzlists of all time, and the puzzle below is adapted from one of his many brilliant creations. Perhaps the biggest shock is that it can be solved without higher mathematics or even a calculator.

The diagram displays three square plots of land. Plot A is 388 square miles, plot B is 153 square miles, and plot C is 61 square miles. How big is the triangular plot of land between A, B, and C? *Answer, page 262*

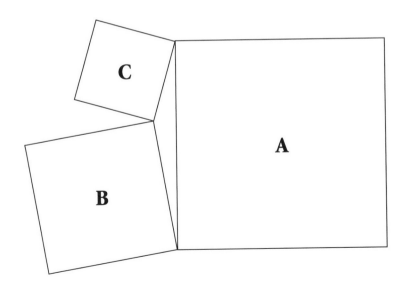

Walking the Blank

The number below is intended to be a 28-digit number, but ten of the digits have been left blank. These blanks are to be filled with the digits 0, 1, 2, 3, 4, 5, 6, 7, 8, and 9—which, for the record, can be done in 10! = 3,556,800 different ways. What is the probability that the resulting 28-digit number will be divisible by 396?

5_383_8_2_936_5_8_203_9_3_76

Answer, page 263

Breaking the Hex

If you start with a regular hexagon and join its vertices as shown, you will create a smaller regular hexagon in the middle. If the larger hexagon has an area of one square foot, what is the area of the smaller hexagon? *Answer, page 264*

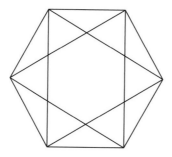

Domino Theory

Suppose you had two dominoes that looked like this:

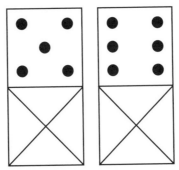

The X's at the bottoms of the dominoes mean that these areas are not visible, so all you know is that one of the dominoes has a five-spot and the other has a six-spot. What is the probability that you can form an end-to-end chain of all 28 dominoes—with the two depicted dominoes in first and last position—subject to the usual rule that the number on the right of any domino in the chain always equals the number on the left of the next domino?

Answer, page 264

The Wayward Three

There is an integer whose first digit is 3 having the property that if you take the 3 from the beginning of the number and place it at the end, you will have multiplied the original number by 3/2. What is that number? *Answer, page 265*

Down to the Wire

Suppose two evenly matched teams play in the World Series. They are so evenly matched that the probability of either team's winning any particular game is precisely 50 percent. Not only that, the teams don't get overconfident or discouraged, so the 50 percent probability doesn't change as the Series goes on.

Under these conditions, it is quite unlikely that either team will engineer a four-game sweep. In fact, it turns out that a sweep is precisely one-half as likely as the Series' ending in five games. What is the likelihood that the Series will go the full seven games?
Answer, page 266

Easy as A, B, C

Can you find three distinct positive integers A, B, and C such that the sum of their reciprocals equals 1?
Answer, page 266

The Beanpot Rally

Every winter in Boston, four area schools—Boston College (BC), Harvard, Northeastern, and Boston University (BU)—compete in a hockey tournament called the Beanpot. One year BU defeated Northeastern in the first round, while Harvard defeated BC. BU went on to defeat Harvard in the final, while BC defeated Northeastern for third place. The tournament draw looked like this:

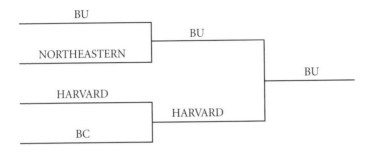

If the above set of results—coupled with the third-place play-off—is considered one possible outcome for the tournament, how many possible outcomes are there in all? *Answer, page 266*

3, 4, 6, Hike!

A hiker went on a little expedition. The first part of the trip was on level ground, and he walked at 4 miles per hour. The second part was uphill, and his speed slowed to 3 m.p.h. He then retraced his steps, going 6 m.p.h. downhill and then the same 4 m.p.h. back to his starting point. Given that the total trip took five hours, how far did the hiker walk? *Answer, page 266*

Even Steven

The one-digit odd numbers—1, 3, 5, 7, and 9—add up to 25, while the one-digit even numbers—0, 2, 4, 6, and 8—add up to 20. Your challenge is to arrange the numbers in such a way that the odd numbers and even numbers have the same value. You can use +, −, ×, ÷, and also combine digits to make multi-digit numbers. *Answer, page 267*

Top Score

If a bunch of positive integers adds up to 20, what is the greatest possible *product* of these numbers? *Answer, page 267*

A Bridge Too Far

In the game of duplicate bridge, the idea is to get a better score than the other pairs who, during the course of the session, play the same hands at different tables. A pair will get one point for every score they beat, and half a point for every score they tie.

One particular hand was played eight times. All pairs playing North-South scored either +450 or +420. (The derivation of these scores doesn't matter, although they are common results for a contract of, say, four hearts.) One of the pairs that scored +420 noted that this score was worth 2½ points at the end of the session. How many points would this pair have received had they scored +450 instead? *Answer, page 267*

Seven-point Landing

Plot seven points on a sheet of paper in such a way that if you choose any three of them, at least two will be precisely one inch apart. (Note: You don't have to bend the paper in any way. The points should all be in the same plane.)

Answer, page 267

'Round Goes the Gossip

One puzzle that is destined to become a classic (and is being mentioned here in order to improve its chances) is the "gossip" puzzle, which has several variations but goes essentially like this:

There are six busybodies in town who like to share information. Whenever one of them calls another, by the end of the conversation they both know everything that the other one knew beforehand. One day, each of the six picks up a juicy piece of gossip. What is the minimum number of phone calls required before all six of them know all six of these tidbits?

Answer, page 268

Hitter's Duel

Let's say that Ty Cobb's season batting average is the same as Shoeless Joe Jackson's at the beginning of a late-season doubleheader. (Assume both players have had hundreds of at bats.) Cobb went 7 for 8 on the day (.875), while Jackson went 9 for 12 (.750).

But at the end of the day, Jackson's season average turned out to be higher than Cobb's. How is this possible?

Answer, page 269

Visible and Divisible

Can you replace the missing digits in the number 789,XYZ so that the resulting number is divisible by 7, 8, and 9? The only restriction is that you cannot use a 7, an 8, or a 9.

Answer, page 269

Scale Drawing

The Celsius scale is derived from the Fahrenheit scale by making a linear adjustment: Specifically, whereas the freezing and boiling points of water are 32 degrees and 212 degrees Fahrenheit, the Celsius scale was created to make these important points 0 degrees and 100 degrees, respectively.

There is one temperature that reads the same on both the Fahrenheit and Celsius scales. What is it?

Answer, page 269

Playing All the Angles

Pictured below is an isosceles right triangle. Using three straight cuts, divide the triangle into four pieces that can be put together to create two smaller isosceles right triangles that are different sizes.

Answer, page 270

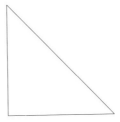

How Big?

The figure below shows two circles of radius one, with the center of each lying on the circumference of the other. What is the area of the wedge-shaped region common to both circles?

Answer, page 270

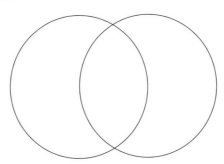

Tunnel Division

A train traveling at 90 miles per hour takes four seconds to completely enter a tunnel, and 40 additional seconds to completely pass through the tunnel. How long is the train? How long is the tunnel?
Answer, page 271

Prime Time

Arrange the digits 0, 1, 2, 3, 4, 5, 6, and 7 so that the sum of any two consecutive digits is a prime number. (Remember that 1 is not considered prime.) There is more than one possible answer.
Answer, page 271

Triangle Equalities

There are precisely three right triangles with integral sides having the property that their area is numerically equal to twice their perimeter. Can you find them?
Answer, page 272

Dividing the Pentagon

It is easy to divide an equilateral triangle into three equal (though not equilateral) triangles. It is even simpler to divide a square into four equal squares. These constructions are given on the next page:

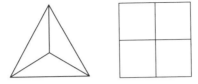

Now for the tough part. Can you divide the regular pentagon below into five equal pentagons? (The smaller pentagons will of course not be regular pentagons.)

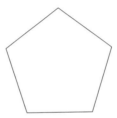

Answer, page 272

Planting the Sod

It is well known that a number is divisible by 9 if and only if the sum of its digits is divisible by 9. For example, $3 + 9 + 6 = 18$ is divisible by 9, and $396 = 9 \times 44$. Let's introduce the SOD operator as one that sums the digits in a number, so that SOD(396) = 18.

With all this in mind, can you determine the value of SOD(SOD(SOD(4444^{4444})))?

One piece of advice: Please don't start calculating 4444^{4444}, as the world will likely end before you or your descendants ever finish the job. *Answer, page 272*

Higher Than You Think

What is the smallest number N such that it is impossible to have $1.00 in change consisting of precisely N coins? You can use half-dollars, quarters, dimes, nickels, and pennies—and even a Sacajawea dollar for the case N = 1.
Answer, page 273

Pocket Change

In the unlikely event that the previous problem was too easy, here's a genuine toughie for you along similar lines:

Suppose a friend of yours announces that he has a number of coins in his pocket that add up to precisely one dollar. When he tells you how many coins he has, you ask if any one of them is a half-dollar, and he answers no. You quickly realize that you can't tell for sure what coins he has, because there are six different combinations that produce precisely one dollar.

How many coins does your friend have in his pocket?
Answer, page 273

Square Not

Note first that the number 150 is expressible as the sum of distinct squares, as shown below.

$$150 = 100 + 49 + 1 = 10^2 + 7^2 + 1^2$$

We'll spare you the trouble, but you can take our word for it that every number greater than 150 is also expressible as the sum of distinct squares. But there are 37 numbers that cannot be expressed in this fashion. Care to find the largest one?

Answer, page 274

Dueling Weathermen

The weathermen at the two TV stations in town—WET and WILD—were locked in a constant battle to come out with the most accurate forecasts. The WET weatherman's long-term accuracy record was $3/4$; the WILD weatherman's long-term accuracy record was $4/5$.

For the day ahead, WET predicted rain, while WILD predicted sun. Assuming that rain and sun were objectively as likely as one another, what were the odds of rain based on the two stations' forecasts? *Answer, page 274*

High Math at the 7–11

A customer brought four items to the cashier of the 7–11 convenience store. "That'll be $7.11," the clerk said. At first the customer thought it was a joke. "Ha, ha. Seven-eleven. I get it." But the clerk was serious. "No, really. I multiplied the prices of the four items you gave me, and I came up with $7.11."

"You *multiplied* them?" the customer asked. "You're supposed to *add* them, you know."

"I know," said the clerk, "but it doesn't make any difference. The total is still $7.11."

What were the prices of the four items?
Answer, page 275

Who Am I?

I am a number with the following properties:

- If I am not a multiple of 4, then I am between 60 and 69.
- If I am a multiple of 3, I am between 50 and 59.
- If I am not a multiple of 6, I am between 70 and 79.

What number am I? *Answer, page 275*

Home on the Range

Imagine a pasture that is just big enough to feed 11 sheep for a total of 8 days. It turns out that if we reduce the number of sheep to 10, they would be able to eat for 9 days.

Theoretically, how long could two sheep last?

Answer, page 276

Heads or Tails

If you flip a coin five times, what is the probability that three or more consecutive flips come out the same?

Answer, page 276

ANSWERS

CHILD'S PLAY?

Sox Unseen (page 21)

Sam has to take out 3 sox; then he's bound to get two of the same color.

Gloves Galore! (page 22)

This is trickier than the sox, because some gloves fit on the right hand and some on the left. You *might* pick out all 12 left hand gloves, one right after the other, but then the next must make a pair; so you need to take 13 gloves to make sure.

Birthday Hugs (page 23)

Each girl hugs 3 others; so it looks like $4 \times 3 = 12$ hugs, but that would be counting Jenny hugging Janey as one hug, and Janey hugging Jenny as another, counting each hug twice. Actually, there are six hugs altogether.

Gold Star answer: Jabberwocky *by Lewis Carroll.*

Mathbit — Fifty/Fifty Chance (page 62)

Only 23! Try it in your class.

Sticky Shakes (page 24)

Same trick as with the hugs. Either you can say each of the 7 shakes with six; so the total is a half of 7 × 6, or 21 shakes. Or you can say John shakes with 6; Jack shakes with 5 others (don't count John again); Jake shakes with 4 others, and so on. The total number of handshakes is 6 + 5 + 4 + 3 + 2 + 1 = 21.

The Wolf, the Goat, and the Cabbage (page 25)

Take the goat across. Go back; take the wolf across, and bring the goat back. Take the cabbage across. Go back for the goat. Then the goat is never alone with either the wolf or the cabbage.

Floating Family (page 26)

The two kids row across. One brings the boat back. Then Mom rows across, and the other kid brings the boat back. Both kids row across. One brings the boat back. Then Dad rows across, and the second kid takes the boat back to collect her brother.

Slippery Slopes (page 27)

Ten days. After 9 days and 9 nights, she is at 9000 feet. On the 10th day she climbs 3000 feet to the summit!

The Long and the Short of the Grass (page 28)

They mowed the grass on 9 Saturdays, earning 9 × $2 = $18, and missed 6 Saturdays, losing 6 × $3 = $18.

Nine Coins (page 29)

She set up the coins as shown. Can you find the rows now? Let's count them.

The 3-coin rows are:
3 rows across (top, middle
and bottom), 4 diagonal (two
one way two the other,
1 down the middle, and…

⭕   ⭕   ⭕

⭕ ⭕ ⭕

⭕   ⭕   ⭕

…2 *long* diagonals criss-crossing through the center!

Tricky Connections (page 30)

No, it can't be done. The connections are impossible without allowing one line to cross another, using a bridge, or putting a line under a house!

Odd Balls (page 31)

Yes, this can be done, but you have to put at least one bag inside of another. You could put 3 balls in each of 3 bags, and then one of these bags inside the fourth bag. Or you could put all the balls in bag 1, put that in bag 2, put that in bag 3, and put the whole lot in bag 4. And there are many others!

Cube of Cheese (page 32)

You have to make one cut to make each face of the cube; so however you pile the pieces, you must make six cuts in all.

Potato Pairs (page 33)

Add all the weights together and divide by two to get the total weight of the three potatoes: $(3+5+4)/2 = 6$ pounds. Now, since A and B together weigh 3 pounds, and A + B + C together weigh 6 pounds, then C must weigh 3 pounds. A and C together weigh 5 pounds, which means that A must weigh 2 pounds; so Cal should buy either A and B or A and C.

Sugar Cubes (page 34)

1. The first trick is to count the zeros! To find out how big the big cube is you need to find the cube root of a million. A million has six zeros—1,000,000—so its cube root must have one third of six—two zeros—100.

The cube root of a million is a hundred. So the big cube is 100 cubes long, 100 wide, and 100 high. Each cube is half an inch; so 100 cubes is 50 inches, or just over 4 feet long. You would not fit this under a table, but it would go easily in a garage.

2. This time you are making a square; so you need the square root of a million. A million has six zeros; its square root must have half six; that is, three zeros—1000. The square root of a million is a thousand. So the big square on the ground is 1000 half inches long and 1000 half inches wide. 1000 half inches is 500 inches; dividing by 12 will give you 41 feet 8 inches. You could fit this square on a tennis court.

3. The pile is a million cubes high; a million half inches, or 500,000 inches. Divide by 12 for 41,666 feet 8 inches. This is higher than Mount Everest. You could make one pile as high as Mount Everest and one as high as Mount Adams, and still have a few cubes left over!

Mathbit — Five Odd Figures (page 70)

1 + 1 + 1 + 13. When you try this on your friends, take care to say "figures" rather than "numbers," or the trick won't work!

Crackers! (page 35)

This is surprisingly easy; the trick is to add a plain cracker. Then Marty has a choice of 2—mayo or plain. Marty and Jake have a choice of 4; when Hank arrives they have a choice of 8, since the number of choices doubles with each new person. So when Hank comes there will be 16 choices—or 15 spreads. When Charlie is there they'll have 32 choices—31 spreads. And Fred will bring the total to 64 choices—63 spreads!

Crate Expectations (page 36)

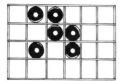

There are many different patterns that work, but here is an easy one to remember.

Now try ten bottles!

Witches' Brew (page 37)

The pan holds 3 pints; fill it and then fill the jug from it. The jug holds 1 pint; so that leaves exactly 2 pints in the pan. Pour it into the cauldron and carry on cooking!

Witches' Stew (page 38)

Fill the pitcher to the brim. Use it to fill the pot, which leaves just 2 pints in the pitcher. Empty the pot back into the bucket. Pour the 2 pints from the pitcher into the pot. Fill the pitcher again. Now carefully top off the pot from the pitcher. This will take exactly 1 pint, because there are 2 pints in it already. That leaves exactly 4 pints in the pitcher—pour them into the cauldron!

The Pizza and the Sword (page 39)

To get the maximum number of pieces you must make sure that each cut crosses all the previous cuts, but not at old crossings. If all three cuts cross in the middle, you can get only 6 pieces, but if you keep the crossings separate you get 7.

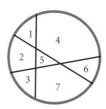

Pencil Squares (page 40)

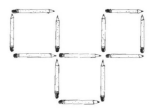

From puzzle, remove top middle pencil and two lower-left corner pencils.

Pencil Triangles (page 41)

What you have to do is build them into a 3-dimentional pyramid, with a triangle on each of its three sides, and *one underneath*.

Cyclomania (page 42)

Donna saw more than one tricycle; so at least six of the wheels must have belonged to tricycles. Suppose there had been three tricycles; then they would have had 9 wheels, which would have left only 3 wheels—not enough for more than one bicycle. So there must have been exactly two tricycles. That makes 6 wheels; so the other 6 wheels must have belonged to bikes; therefore there must have been 3 bikes and 2 trikes.

Spring Flowers (page 43)

The way to figure this out is to start at the end with 39 petals, and remember that the primroses must provide either 5 petals or 10 or 15 or 20 or 25 or 30 or 35—a multiple of 5. Now suppose there was only one celandine (8 petals); that would leave 31 petals ($39 - 8 = 31$). The rest can't be primroses, because 31 is not an exact multiple of 5. Suppose there were two celandines; that would make 16 petals, leaving 23. No good! Three celandines $= 24$ petals, and $39 - 24 = 15$ petals. Bingo! The answer must be 3 celandines and 3 primroses. Check: $3 \times 8 = 24$ and $3 \times 5 = 15$, and $24 + 15 = 39$. So Rosa is $(3+3)$, or 6.

Sesquipedalian Farm (page 44)

If 1½ hens lay 1½ eggs in a day and a half,
then 1 hen lays 1 egg in a day and a half,
and 2 hens lay 2 eggs in a day and a half,
and 7 hens lay 7 eggs in a day and a half,

so 7 hens lay 14 eggs in three days,
so 7 hens lay 28 eggs in six days,
so 7 hens lay 42 eggs in nine days;
add what they lay in 1½ days
($+7$ eggs in a day and a half),
so 7 hens lay 49 eggs in 10½ days, which is a week and a half.

Three-Quarters Ranch (page 45)

We know 1¾ ducks lay 1¾ eggs in 1¾ days. Imagine that we can replace 1¾ ducks by a special bird called a Megaduck. Then we can say:

1 Megaduck lays 1¾ eggs in 1¾ days;
so 1 Megaduck lays 1 egg in 1 day;
so 1 Megaduck lays 7 eggs in 1 week.

Now, how many Megaducks are there in 7 ducks?
Answer 7 divided by 1¾, which turns out to be 4!
($4 \times 1 = 4$, and $4 \times \frac{3}{4} = 3$, and $4 + 3 = 7$).

The original question was, "How many eggs do 7 ducks lay in a week?"
Since 7 ducks is the same as 4 Megaducks, and since each Megaduck lays
7 eggs in a week, 4 Megaducks lay $4 \times 7 = 28$ eggs a week. So, 7 ducks
lay 28 eggs a week.

Cookie Jars (page 46)

Joe has no cookies; so this puzzle is easy. If Ken gave him one, he'd have
a total of one; so if they have the same number, Ken must also have one
left. Therefore Ken must have two to begin with.

Fleabags (page 47)

Captain has two fleas; Champ has four.

The Rolling Quarter (page 48)

Twice. Try it and see.

Sliding Quarters (page 49)

Pick up and use the dark quarter to move the other ones.

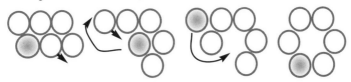

Practice this before you try to show anyone!

Picnic Mystery (page 50)

The cake isn't in the CAKE box, and it can't be in the COOKIE box, because that's the only place for the fruit; so it must be in the box labeled FRUIT.

Find the Gold (page 51)

Ask to see a sample from the box labeled MIXTURE, because you know it isn't one. If it's gold, that box must be full of gold; take it. If the sample is of iron, then take the box labeled IRON, because you know the gold is in neither the GOLD nor the MIXTURE.

Frisky Frogs (page 52)

Freda steps, Fred hops over her, Frank steps, Freda hops, Francine hops, Fergie steps, Fred hops, Frank hops, Frambo hops, Freda steps, Francine hops, Fergie hops, Frank steps, Frambo hops, Fergie steps—and they are all across, in 15 moves.

Leaping Lizards (page 53)

Use these rules as a guide: 1. Don't move a boy next to another boy—or a girl next to another girl—until you reach the other side; 2. Step if you can, hop if you can't step; 3. Once you have moved a girl, keep moving girls till you have to stop; then move boys till you have to stop. The quickest has 23 leaps.

Chewed Calculator (page 54)

To make 3: $4+4+4=/4=3$
To make 4: $4-4+4\mathrm{sqrt}=+\mathrm{sqrt}=4$
To make 5: $4{*}4=+4=/4=5$
To make 6: $4+4+4=/4\mathrm{sqrt}=6$

To make 7: $-4/4=+4+4=7$
To make 8: $4+4+4-4=8$
To make 9: $4/4+4+4=9$
To make 10: $4+4+4-4 \ \mathrm{sqrt}=10$

Crushed Calculator (page 55)

To make 1: $1 \times 2 = +3-4 = 1$

To make 2: $4-3 = \times 2 \times 1 = 2$

To make 3: $4-2-1 = \times 3 = 3$

To make 4: $4+3-2-1 = 4$

To make 5: $4+3-2 \times 1 = 5$

To make 6: $4+3-2+1 = 6$

To make 7: $2-1 = \times 4+3 = 7$

To make 8: $1-2+3 = \times 4 = 8$

To make 9: $4 \times 3 = -2-1 = 9$

To make 10: $4 \times 2 = +3-1 = 10$

…and see how much further you can go!

Squares & Cubes (page 56)

$64 = 8 \times 8$ and $4 \times 4 \times 4$

Cubes & Squares (page 57)

The only number between 100 and 999 that is both a square and a cube is 729, which is 27×27 and $9 \times 9 \times 9$.

Old MacDonald (page 58)

All the 12 wings must have belonged to turkeys, because pigs don't usually have any; so he must have had 6 turkeys (with 2 wings each). The 6 turkeys must have had 12 legs; leaving 12 legs for the pigs, and since each pig has 4 legs, that makes 3 pigs. So Old MacDonald had 3 pigs and 6 turkeys.

Old Mrs. MacDonald (page 59)

Mrs MacDonald counted 12 heads, so she must have had 12 animals. If they had all been chickens she would have had 24 legs; if they had all been cows she would have had 48 legs. The difference between these two is 24, or 2 legs more than 24 for each cow. She counted 34 legs. That is 10 more than 24; so she must have had 5 cows.

Check: 5 cows = 5 heads; 7 chickens = 7 heads;
total 5+7= 12 heads.

And 5 cows = 20 legs; 7 chickens = 14 legs;
total 20+14 = 34 legs.

Wiener Triangles (page 60)

You can make 5 triangles, including the big one around the outside.

Tennis Tournament (page 61)

In a knockout tournament, every player has to lose one match—except
the winner, who loses none. So the total number of matches is one less
than the number of players. If 27 enter, there will be 26 matches.

House Colors (page 62)

Houses #1 and #3 must be green, because they are not next door to one
another; so Bernice and her family live in the middle house.

LoadsaLegs (page 63)

One had 6 legs; the other had 10.

Antennas (page 64)

Abot has 4 antennas; Bbot has 8.

The Power of Seven (page 65)

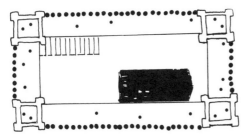

After 4 have been killed, and there are 20 left, the commander must put 2 in each corner tower, and 3 along each side wall.

The Power of Seven Continues (page 66)

Yes, they can survive the final attack. Four defenders go in one corner tower, three in all the others. 15 men make 7 on each side!

Bundles of Tubes (page 67)

These are the shapes formed by bundles of tubes naturally, starting with 1 and then making hexagons of 6, 12, and 18:

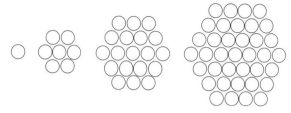

1 tube + 6 = 7 tubes   + 12 = 19 tubes   + 18 = 37 tubes

Pyramids (page 68)

Susie needs 56 cookies; Ben needs 55.

Wrong Envelope? (page 69)

Think of the three envelopes as A, B, and C, and the three letters as a, b, and c. Then you can write down the six different ways of arranging the letters like this:

1	2	3	4	5	6
Aa	**Aa**	Ab	Ab	Ac	Ac
Bb	Bc	Ba	Bc	Ba	**Bb**
Cc	Cb	**Cc**	Ca	Cb	Ca

Only #1 has all the letters in the right envelopes; so there are five ways of putting at least one letter into the wrong envelope, and your chance of getting it all right just by luck is one in six.

There are only two arrangements with all the letters in the wrong envelopes. Can you find any arrangements with two letters in the right envelopes and one wrong?

Squares & Cubes & Squares (page 70)

1) 26 is one more than 25 (5×5) and one less than 27 (3×3×3).
2) 123 is two more than 11×11 (121) and two less than 125 (5×5×5).

Good Neighbor Policy (page 70)

The division of the pots of jam was simple; the neighbors were father, son, and grandson, and so they got one pot each.

Cutting the Horseshoe (page 71)

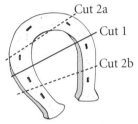

Cut 2a

Cut 1

Cut 2b

First cut across both arms, leaving two holes below the cut on each side, to give three pieces. (Cut 1)

Pile these up so that your second cut snips each of the bottom ends in half, and cuts out the top section with one hole. (Cut 2)

You have seven holey pieces!

Multisox (page 72)

One has 14 feet, the other 22, making 36 in all.

Three Js (page 73)

Joan is 32; Jane is 28; Jean is 16.

Architect Art (page 74)

Start at one of the top right corners, and you will have no trouble.

No Burglars! (page 75)

Start in the kitchen, and go first through either door #6 or door #10; then you will find the rest easy.

Train Crash (page 76)

In the hour before the crash one train must have traveled 25 miles, and the other 15. So one hour before the crash they were (25 + 15) or 40 miles apart.

Puzzle of the Sphinx (page 77)

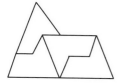

Perforation! (page 78)

Tear the sheet in a zigzag pattern, starting from the left one stamp up from the bottom. Then slide the lower piece up to the right.

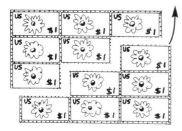

Disappearing Apples (page 79)

He had six apples to start with, and ate two the first day and two the second.

Colored Balls #1 (page 80)   ## Colored Balls #2 (page 81)

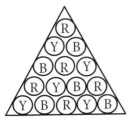   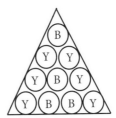

Logical Pop (page 82)

You should shake either the can labeled MILK SHAKE or the can labeled POP. The potato chips must be in one of those two cans, and they should rattle.

Suppose you shake the MILK SHAKE can; if it rattles it contains the chips; the pop must be in the can labeled POTATO CHIPS.

If it doesn't rattle, it must contain the pop.

Meanwhile if you shake the POP can and it rattles, the chips must be in it; so the milk shake must be in the can labeled POTATO CHIPS, which means the pop must be in the MILK SHAKE can. If it doesn't rattle, the pop must be in the can labeled POTATO CHIPS.

Mathbit — Four 9s (page 42)

99⁹⁄₉.

WORKING TOWARDS
WIZARDRY

Secret Number Codes (page 85)

For each letter look at the number in the table, and add 1. So A=2, B=3, C=4 etc. The message reads meet me at seven.

Shape Code (page 86)

The message reads phone me soon.

Magic Triangle (page 88)

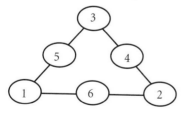

Now can you rearrange the same numbers so that each side totals 12?

Magic Hexagon (page 89)

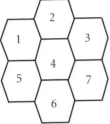

Make sure the middle number, in this case 4, is in the center.

Easier by the Dozen (page 90)

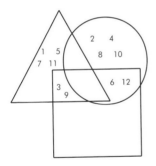

Who Is Faster? (page 91)

Hector can run a mile in eight minutes, so it takes him 64 minutes to run eight miles. But Darius can run eight miles in just 60 minutes, so Darius is faster. There could be another question here. Could Hector keep up his eight-minute pace for an entire hour? Maybe, maybe not. However, if he can't keep it up, it proves that Darius is even faster!

The Average Student (page 92)

Three five-star homework papers will do the trick. Altogether they account for $3 \times 5 = 15$ stars. Adding the single star from the first homework gives 16 stars from four assignments, for an average of four stars per assignment.

Looking at the problem another way, note that the one-star homework paper was three stars under the desired average of four stars. Each five-star homework gains one point on the average, so it takes three of them to balance things out.

Big Difference (page 93)

The biggest possible difference is as follows:

$$
\begin{array}{ccc}
 & \boxed{7} & \boxed{6} \\
- & \boxed{2} & \boxed{4} \\
\hline
 & 5 & 2
\end{array}
$$

Not Such a Big Difference (page 93)

If we want the smallest possible difference instead, that would be

$$
\begin{array}{ccc}
 & \boxed{7} & \boxed{2} \\
- & \boxed{6} & \boxed{4} \\
\hline
 & & 8
\end{array}
$$

ANSWERS
Working Towards Wizardry

Square Route (page 94)

The Missing Six (page 94)

There is more than one answer to this puzzle.

$$(2) + (5) = (7)$$

$$(4) - (3) = (1)$$

Here is another:

$$(2) + (5) = (7)$$

$$(4) - (3) = (1)$$

Donut Try This at Home (page 95)

Suppose a regular donut has 100 calories. If a low-calorie donut has 95 percent fewer calories, it must have 5 calories. Therefore you must eat 20 low-calorie donuts to get as many calories as you get from one regular donut.

The Long Way Around (page 96)

The total distance around the diagram is 8 + 15 + 8 + 15 = 46 units. It makes no difference that the upper right portion of the diagram includes a set of "steps." That's because if you pushed out those steps, you could create an 8-by-15 rectangle, and the distance around that rectangle (its "perimeter") would be precisely 46 units.

A Very Good Year (page 97)

The next year to have the same property will be 2307: 23 + 07 = 30.

Long Division (page 98)

Cover up the left half of the figure 8 below. What's left looks a lot like a 3, doesn't it?

When in Rome (page 99)

Now cover up the bottom half of the figure below, which happens to be 9, in Roman numerals. What's left should give you 4, also in Roman numerals.

IX

Diamond in the Rough (page 100)

The only diamond that is not symmetrical is the seven of diamonds.

Three's a Charm (page 101)

The item costs 17 cents. To purchase it requires four coins: one dime, one nickel, and two pennies. To purchase two items (34 cents) requires six coins: one quarter, one nickel, and four pennies. To purchase three items (51 cents) requires only two coins: one half-dollar and one penny.

Who Is the Liar? (page 102)

Daniel is the liar. To see why, we examine one case at a time, using the fact that only one person is lying.

If Andrew were lying, the number would have three digits. (It couldn't have just one digit, because then it couldn't be divisible by 25, and Daniel would also be lying.) But if the number had three digits, either Barbara or Cindy would have to be lying, because 150 is the only three-digit number that goes evenly into 150. Therefore Andrew must be telling the truth, because there can only be one liar.

If Barbara were lying, then the number does not go into 150. But then either Andrew or Daniel must be lying, because the only two-digit numbers that are divisible by 25 (25, 50, and 75) all go evenly into 150. So Barbara must be telling the truth.

If Cindy were lying, then the number would be 150. But then Andrew would also have to be lying, because 150 has three digits, not two. And we know Andrew is telling the truth.

So, the only possibility left is that Daniel is the liar, and this works out. If the number were 10, for example, Daniel would be lying, but the other three statements would all be true.

The Powers of Four (page 102)

Ernie came up with the number 1,048,576. Note that all of Bert's numbers end in 4, while all of Ernie's numbers end in 6. That's all you need to know!

High-Speed Copying (page 103)

Eight copiers can process 800 sheets in 4 hours. Doubling the number of copiers will double the output without changing the amount of time required.

Divide and Conquer (page 104)

```
        5  4
  9  4  8  6
   - 4  5
  ----------
        3  6
      - 3  6
  ----------
           0
```

Agent 86 (page 105)

32	19	27	8
10	25	17	34
9	26	18	33
35	16	24	11

Comic Relief (page 105)

One 10-ruble book, two 2-ruble books, and three 1-ruble books add up to six books and 17 rubles.

Pieces of Eight (page 106)

Each point from A to H (the "vertices" of the octagon) can be connected with five other points to form a diagonal. That seems to make a total of 8 × 5, or 40 diagonals. However, as it said in the hint, the diagonal from A to E is the same as the diagonal from E to A, and you can't double-count. You need to divide 40 by 2 to get the actual answer—20 diagonals.

On the Trail (page 107)

1,009 in Roman numerals is MIX. It is the only number that is a common English word in Roman numerals.

From Start to Finish (page 107)

The total number of trips from S (Start) to F (Finish) equals 10. The easiest way to solve the problem is to use the diagram and count them up!

For a more systematic way to arrive at the answer, we start by observing that at each point you have a choice between moving across or moving down. Altogether you have to move across three times and down twice in order to get from S to F. So a pattern that would get you from S to F might look something like AADAD, where A means across and D means down.

Using our notation of A's and D's, the ten patterns are as follows:

AAADD	ADAAD
AADAD	DDAAA
AADDA	DADAA
ADDAA	DAAAD
ADADA	DAADA

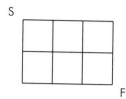

Apple Picking (page 108)

They all cost the same!
Ten apples = 1 bag (5¢) plus 3 apples at 15¢ each (45¢) = 50¢
Thirty apples = 4 bags (20¢) plus 2 apples at 15¢ each (30¢) = 50¢
Fifty apples = 7 bags (35¢) plus 1 apple at 15¢ (15¢) = 50¢

Playing the Triangle (page 109)

The key to this puzzle is that if you add up the lengths of any two sides of any triangle, the sum must be greater than the third side. Why is this true? Because the shortest distance between any two points is a straight line. For example, in the diagram below, AB + BC could never be less than AC, because then the indirect route from A to C—stopping off at B along the way—would be shorter than the direct route!

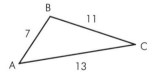

What this means is that 5 cannot be the third side, because 5 + 7 < 13. In the same way, 21 is impossible, because 7 + 13 < 21. That leaves 11 as the only possible answer.

Generation Gap (page 109)

Grandpa Jones is 78. His four grandchildren are 18, 19, 20, and 21. Note that 18 + 19 + 20 + 21 = 78.

It is not possible for the sum of four consecutive numbers to be equal to 76 or 80. In general, the sum of four consecutive numbers will never be divisible by 4! (Both 76 and 80 *are* divisible by 4.)

The Birthday Surprise (page 110)

The professor had forgotten that the class included a pair of identical twins! (This is apparently a true story, the professor in question being famed logician and author Raymond Smullyan.)

The Run-Off (page 111)

Burt will win. Why? Because when the two of them ran against Alex, Burt was exactly 20 meters ahead of Carl at the moment Alex crossed the finish line. Therefore, if Burt were to give Carl a 20-meter head start, they will be even at that point. But that point is 20 meters from the finish line, and Burt is the faster runner. Therefore he would win the race—but not by much!

The French Connection (page 112)

One way to compute the average test scores for the two students is to add up their individual test scores and divide by 5.

Average for Sandy = (94+79+84+75+88)/5 = 420/5 = 84
Average for Jason = (72+85+76+81+91)/5 = 405/5 = 81
Sandy has a three point advantage.

An easier way might be to arrange the test scores in the following way:

Sandy: 94 88 84 79 75
Jason: 91 85 81 76 72

It is now easy to see that Sandy has a three-point advantage the whole way through, so her average must be three points higher.

Mirror Time (page 113)

The answer is 49 minutes, the time between 12:12 and 1:01.

Staying in Shape (page 113)

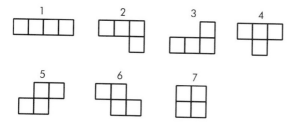

Going Crackers (page 114)

The junior employee argued that what the survey said was 1) Crackers are better than nothing, and 2) Nothing is better than peanuts. Putting the two together, you get that crackers are better than peanuts!

The Missing Shekel (page 115)

The problem with the price of five rutabagas for two shekels is that the five rutabagas consist of 3 cheap ones (the three for a shekel variety) and 2 expensive ones (the neighbor's two for a shekel batch). By selling all 30 rutabagas in this manner, the farmer is basically selling 20 at his price and 10 at his neighbor's more expensive price—not 15 at each price. That's why he ends up a shekel short.

Quarter Horses (page 116)

B is shortest, A is in the middle, and C is longest. Note that A is just the radius of the circle, while B is clearly shorter than the radius. As for C, you can see from the picture below that the fence is longer than the radius. So, if C is longer than A, and B is shorter than A, you have your answer: B is the shortest and C is the longest.

GREAT MATH CHALLENGES

A Roman Knows (page 184)

Because $X \times X < 10$, it follows that $X = 1$, 2, or 3. However, $X = 1$ is obviously impossible, because then XLIV would have to equal CDXL. It can also be shown that $X = 3$ doesn't work: L would then have to equal 1 or 2, and either choice produces a contradiction. If $L = 1$ then $V = 7$, but then I would also have to equal 7 to produce the X in CDXL. Similarly, if $L = 2$ then $V = 4$ but I would also have to be 4.

It follows that $X = 2$, and it turns out that there are two solutions: $2864 \times 2 = 5728$ and $2814 \times 2 = 5628$.

Page Boy (page 184)

The sum must have been 412, because $412 = 48 + 49 + 50 + 51 + 52 + 53 + 54 + 55$.

512, on the other hand, is a power of two, and no power of two can be expressed as the sum of consecutive integers. The proof of this follows:

Note that the sum of the first n integers is given by $\frac{n(n+1)}{2}$. The sum of consecutive integers beginning with $m + 1$ and ending with n is given by $\frac{n(n+1)}{2} - \frac{m(m+1)}{2} = \frac{(n^2 - m^2) + (n - m)}{2}$.

But $n^2 - m^2 = (n + m)(n - m)$, so this expression becomes $(n + m + 1)(n - m)/2$. Now note that $(n + m + 1)$ and $(n - m)$ have different parities—meaning that if one is even, the other must be odd. Therefore their product cannot be a power of two, which completes the proof.

Circular Logic (page 183)

Note that the distances XY and YZ must each be less than the diameter of the circle. Also, segments XC and CZ must sum to precisely the diameter of the circle, because each one of these two segments is a radius.

Therefore, the sum of all four of these segments must be less than three times the diameter of the circle. But the circumference equals π times the diameter, and π is greater than 3, so the family member who walks along the straight paths arrives home first.

No Calculators, Please (page 185)

First note that $(\frac{1}{2}) \times (\frac{2}{3}) \times (\frac{3}{4}) \times (\frac{4}{5}) \times (\frac{5}{6})\ldots \times (\frac{97}{98}) \times (\frac{98}{99}) \times (\frac{99}{100}) = \frac{1}{100}$, which is readily seen because everything other than the initial 1 and the trailing 100 will cancel out.

Therefore the following two numbers multiply to $\frac{1}{100}$:

$(\frac{1}{2}) \times (\frac{3}{4}) \times (\frac{5}{6}) \times \ldots (\frac{97}{98}) \times (\frac{99}{100})$ and
$(\frac{2}{3}) \times (\frac{4}{5}) \times (\frac{6}{7}) \times \ldots (\frac{98}{99}) \times 1$

The top number is the one we are interested in. Clearly it is less than the bottom number, because it is less on a term-by-term basis— $\frac{1}{2} < \frac{2}{3}$, $\frac{3}{4} < \frac{4}{5}$, etc. The top number must therefore be less than the square root of $\frac{1}{100}$, or $\frac{1}{10}$.

Born Under a Bad Sign (page 185)

The answer is E), Sue's 50th birthday.

Clearly Sue's birthday will always be on the 13th, so the only question is when it will be on a Friday. The years that produce a Friday birthday (the same logic holds for any day of the week) follow a cycle of 6, 11, 6, and 5 years: Note that $6 + 11 + 6 + 5 = 28$, and after 28 years the cycle repeats itself. (There are 4 years between leap years and 7 days in the week, so calendars repeat themselves every 4×7, or 28, years.)

To see how the 6, 11, 6, 5 sequence comes up, suppose you were born on a Friday in March of a leap year. The following year your birthday will fall on a Saturday, because 365 has a remainder of 1 upon division by 7. The year after that your birthday will be on a Sunday, then Monday, Wednesday (the key step—skipping a day because this year in the sequence is a leap year), then Thursday and, finally, Friday. In other words, in 6 years your birthday will again be on a Friday, and you'll be precisely between two leap years. Continuing, the next Friday birthday will occur in 11 years, because the 3 leap years during this period will push your birthday up by a total of $11 + 3 = 14$ days, or precisely two weeks. The remainder of the sequence is found the same way, at which point you're back to a leap year, and you start all over again.

We don't know when in the cycle Sue was born, but we can still answer the problem by looking at the sequence 6, 11, 6, 5, 6, 11, 6, 5, 6, 11, 6, 5, 6, 11, 6, 5, 6, 11, 6, 5, and trying to find consecutive entries that sum up to a multiple of 10—10, 20, 30, 40, 50, or 60.

The only one of these multiples of 10 that works is 50, which equals $5 + 6 + 11 + 6 + 5 + 6 + 11$. Specifically, if Sue was born on Friday the 13th during a year prior to a leap year (such as 1987), her 50th birthday will also fall on Friday the 13th. None of the other birthdays listed will ever fall on a Friday, no matter what year she was born in.

One, Two, Three (page 186)

There are four solutions, as follows:

1	2	3	4
192	219	273	327
384	438	546	654
576	657	819	981

Don't Make My Brown Eyes Blue (page 186)

The surprising answer is that everyone with blue eyes vacated the island the morning after the tenth day following the decree.

To see why this is so, suppose that only one person on the island had blue eyes. (Recall that at least one person had blue eyes. This condition may have seemed unimportant at the time, but it was essential to get the proof rolling. This is true of all proofs based on "mathematical induction.") Anyway, after one day, that person, having seen everyone else on the island, would have been able to conclude that he or she had blue eyes, and would therefore leave.

Similarly, if two people had blue eyes, they would see each other on the first day. Then, on the second day, each would see the other, and each would then realize that if that other person was the only one on the island with blue eyes, that other person would have left after the first day. Accordingly, the two would independently realize that they each had blue eyes, and would leave after the second day. And so on, and so on.

The Stamp Collection (page 187)

We know right away that the number of stamps in the collection is divisible by 35 since the number of stamps can be divided evenly by 5 (a fifth of the stamps are in Book 1) and 7 (some number of sevenths are in Book 2) and because 5 and 7 have no common factor. Suppose there are

$^x/_7$ of the stamps in the second book. Together, the first and second books contain $1/5 + {}^x/_7 = {}^{(7 + 5x)}/_{35}$ of the collection. The third book therefore contains $^{(28 - 5x)}/_{35}$ of the collection. If there are C stamps in all, then $35 \times 303 = C \times (28 - 5x)$. But 35 divides into C, so $(28 - 5x)$, which is a positive integer, must equal one of the factors of 303: 1, 3, 101 or 303. Try each of these cases. The only one that leads to a positive integer less than 7 is $(28 - 5x) = 3$, giving $x = 5$. This yields that 303 stamps amounts to $^3/_{35}$ of the collection, so the entire collection equals $35 \times 101 = 3,535$ stamps.

Too Close to Call (page 187)

Take the thirtieth power of both numbers, so as to remove the radical signs altogether. We get the following:

$$\sqrt[10]{10}^{30} = 10^3 = 1000$$
$$\sqrt[3]{2}^{30} = 2^{10} = 1024$$

Therefore the cube root of 2 is slightly larger. (Note that the "K" in the computer memory context is actually 1,024—a power of two, befitting the binary code—but its proximity to 1,000 causes it to be used interchangeably, as in Y2K, etc.)

The Long String (page 187)

The answer is an emphatic yes. The 1,000,000-term sequence 1,000,001 + 2, 1,000,001 + 3, all the way up to 1,000,001 + 1,000,001 consists entirely of composite numbers, because 1,000,001 + K is always evenly divisible by K for any K in this range. More generally, no matter how big the number N is, it is always possible to find N consecutive composite numbers.

First Class Letters (page 188)

The only number between 1 and 5,000 that is alone in its letter class is 3,000. THREE THOUSAND requires 13 letters to write out, and is the *only* number within that range to require precisely 13 letters.

Although we won't provide an exhaustive proof, you would begin by noting duplications among the first nine digits, as in ONE-TWO-SIX, THREE-SEVEN-EIGHT, FOUR-FIVE-NINE. Then we would find duplications among TWENTY and THIRTY, and so on, which would rule out solutions with two digits. The three-digit case is similar. In fact, there would be duplications in the four-digit category as well, except that we restrict ourselves to the 1–5,000 range: Note that 7,000 is also in letter class 13.

Most Valuable Puzzle (page 188)

1) Seven first-place votes would do the trick. Whoever won seven first-place votes would have a minimum of $(7 \times 3) + 3 = 24$ points. The best anyone else could do would be to win 3 first-place votes and 7 second-place votes, for a total of $(3 \times 3) + (7 \times 2) = 23$ points.

2) In theory, the MVP award could be won by someone with only one first-place vote, as long as that person received a second-place vote on each of the remaining 9 ballots. The voting would have to look like this:

Player	1st	2nd	3rd	Points
A	1	9	0	$(1 \times 3) + (9 \times 2) + (0 \times 3) = 21$
B	5	0	5	$(5 \times 3) + (0 \times 2) + (5 \times 1) = 20$
C	4	1	5	$(4 \times 3) + (1 \times 2) + (5 \times 1) = 19$

Performance Anxiety (page 189)

The basis of the misleading arithmetic is the confusion between a percentage increase and a percentage-point increase. The actual performance of the firm's selections, in percent, was a number 32 percent bigger than 28, namely, 28×1.32, or about 37 percent. Not too shabby, but a far cry from a return of 60 percent, which would be a 32 percentage-point improvement over the S&P 500.

Double Trouble (page 189)

$58 \times 3 = 174 = 29 \times 6$

Cereal Serial (page 189)

The answer is $8\frac{1}{3}$ boxes. To see why, note that one of the prizes comes from the first box we buy. The likelihood of getting a new prize from the next box equals $\frac{3}{4}$; on average, therefore, we would need to buy $\frac{4}{3}$ boxes to get a new prize. Proceeding in this same fashion, the third new prize would require an additional $1/(\frac{1}{2}) = 2$ boxes. The fourth would require, on average, an additional 4 boxes. In total, the average number of boxes equals $1 + \frac{4}{3} + 2 + 4 = 8\frac{1}{3}$ boxes.

In real life, of course, you can't buy $\frac{1}{3}$ of a box, but that is still the average number of boxes you'd have to purchase.

Pairing Off (page 190)

The next pair (x,y) satisfying $x^2 - 2y^2 = 1$ is $(17,12)$, and it is easy to check that $17^2 - 2(12^2) = 289 - 2(144) = 1$.

Believe it or not, there is a systematic (though hardly obvious) way of generating all solutions to the equation. If you consider the quantity

$(3 + 2\sqrt{2})$, you will discover that $(3 + 2\sqrt{2})^2 = 17 + 12\sqrt{2}$. If you continued taking successive powers and examining the coefficients, you would discover that the next ordered pair higher than $(17, 12)$ is $(99, 70)$, and so on.

A Square Deal (page 190)

The proper dimensions are as shown. There is no solution if the smallest square is only one mile on each side. There is, however, a solution when the smallest square has a side of two. The plots of the second and third child each measure 96 square miles, and the total area of the father's plot was 196 square miles.

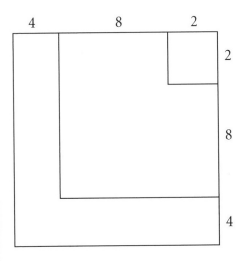

Survival of the Splittest (page 191)

This is one of those problems where it's just as easy to assume that the probability of a successful split is a variable p, rather than a specific number such as ¾. So let p be the probability of a successful split, and P be the probability that the amoeba chain will go on forever.

Look at the second generation of amoebas. With probability p, there are two amoebas in that generation. The probability that these two will generate an infinite chain is $1 - (1 - P)^2$, because $(1 - P)^2$ is the probability that *neither* will do so. Therefore, $P = p(1 - (1 - P)^2)$, because both sides of the equation represent the probability of long-term survival.

Simplifying, we get the equation $pP^2 + (1 - 2p)P = 0$, or $P(pP + (1 - 2p)) = 0$. We assume that P does not equal 0, so that $pP + (1 - 2p) = 0$, or $P = {}^{(2p - 1)}\!/_p$.

Setting p = ¾, we see that $P = (2(¾) - 1) ÷ (¾) = ⅔$.

Shutting the Eye (page 192)

The radius of the big circle is five times the radius of the smallest circle. To see why, consider the diagram below:

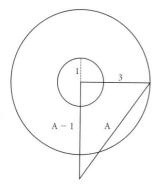

Let's say that the radius of the small circle is 1, and so the radius of the mid-sized circle is 3. If A is the radius of the big circle, then we see that $3^2 + (A - 1)^2 = A^2$. This equation simplifies to $9 - 2A + 1 = 0$, or $A = 5$. (You could also see that the indicated equation yields the Pythagorean relationship for the famous 3–4–5 right triangle. However you slice it, $A = 5$, so the radius of the largest circle is five times that of the smallest one.)

What's in a Name? (page 192)

Of the three cryptarithms, only the middle one has a solution. Here's why.

$$
\begin{array}{cccc}
 & M & A & R & Y \\
+ & T & Y & L & E & R \\
\hline
 & M & O & O & R & E \\
\end{array}
$$

Sorry, Mary, but your equation cannot be solved. Look at the second column. In order for R + E to equal R, E must be either 0 or 9: If E = 0, that means there is no carrying from the first column; if E = 9, there must be a 1 carried over from the first column. Unfortunately, if E = 0 in the first column, there would be a 1 carried over, while if E = 9, there would be nothing carried over—precisely the opposite of our requirements for column two! The addition is therefore impossible.

$$
\begin{array}{ccccc}
 & J & O & Y & C & E \\
+ & C & A & R & O & L \\
\hline
 & O & A & T & E & S \\
\end{array}
$$

There are many solutions to this one. One solution is given by the following:

$$
\begin{array}{ccccc}
  & 5 & 9 & 8 & 3 & 2 \\
+ & 3 & 7 & 1 & 9 & 4 \\
\hline
  & 9 & 7 & 0 & 2 & 6 \\
\end{array}
$$

Observe that the solution is not unique. In particular, you can create another solution by simply interchanging the 1 and the 8 in the middle column.

$$
\begin{array}{ccccc}
  & J & A & M & E & S \\
- &   & E & A & R & L \\
\hline
  & J & O & N & E & S \\
\end{array}
$$

This one has no solution, and to see why you don't need to look beyond the trailing letters. It is impossible for ES − RL to equal ES unless both R and L are zero. But two different letters cannot be given the same number, so no solution is possible.

Thumbs Down (page 193)

The key is to notice that a must be even. But it can't be 4 or 6, because the product would exceed three digits. So a must equal 2. With a little trial and error, the other digits fall into place. The only solution is $275 \times 2 = 572$ in base 8.

Ticket to Ride (page 194)

If F = the number of former stations and N = the number of new ones, we must have $2FN + N(N − 1) = 52$. This equation becomes $N^2 + (2F − 1)N − 52 = 0$. We know that −52 is the product of the roots of this equation, and since the only possible factorizations of 52 are 1×52, 2×26, and 4×13, the sum of the roots must equal 51, −51, 24, −24, 9, or −9. But this sum must equal $2F − 1$, with F a positive integer, so the only possibility is $2F - 1 = 9$, in which case $F = 5$. The equation now factors as $(N + 13)(N - 4) = 0$, and the positive root is $N = 4$. (The possibility $2F - 1 = 51$ yields $F = 26$ and $N = 1$, but the problem stated that several new stations—plural—were added, implying that $N > 1$.) So there were 5 stations originally, and 4 were added.

Going Off on a Tangent (page 194)

Any two tangents to a circle must have equal length, so we can segment the lengths of the diagram as follows:

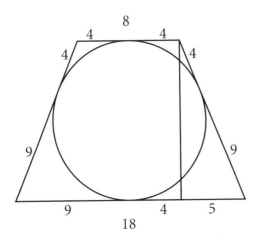

The "5" appears because the length of that segment is the difference of the lengths of two other segments, now known to be 9 and 4, respectively. But the diameter of the circle can now be seen to be a leg of a right triangle with hypotenuse 13 ($= 9 + 4$) and other leg 5, so by the Pythagorean theorem the diameter must be 12.

Surprise Ending (page 195)

The expression $1^n + 2^n + 3^n + 4^n$ is divisible by 5 unless n is divisible by 4. The most straightforward method of proving this result uses modular arithmetic. We start with the following table:

n	1^n	2^n	3^n	4^n
1	1	2	3	4
2	1	4	9	16
3	1	8	27	64
4	1	16	81	256
5	1	32	243	1024

Now we take the remainders upon division by 5, and add those remainders up:

n	1n	2n	3n	4n	1n + 2n + 3n + 4n
1	1	2	3	4	10
2	1	4	4	1	10
3	1	3	2	4	10
4	1	1	1	1	4
5	1	2	3	4	10

Note that n = 4 produces the same results as n = 0 would have, and then the pattern repeats itself. The result is surprising in that a question about divisibility by one number (5) turns out to be related to divisibility by a relatively prime number (4).

Thanksgiving Feast (page 195)

Let T denote the set of pilgrims who had turkey, C the set who had corn, P the pumpkin pie eaters, and M the mince pie eaters. Similarly, let T', C', P', and M' denote the set of people who did not have the indicated dishes—i.e., the complements of the respective sets. Then T' consisted of 10 percent of the pilgrim population, C' 20 percent, P' 30 percent and

M' 40 percent. Because no one had all four dishes, the union of these sets is the entire pilgrim population. Also, since 10 + 20 + 30 + 40 equals precisely 100, these four sets must be disjoint. In particular, that means that P' and M' have no members in common, so everyone must have had one of the two desserts.

All About Pythagoras (page 195)

1) The only numbers that cannot be part of a Pythagorean triple are 1 and 2: In other words, there are no two perfect squares that differ by either 1 or 4. To see why any other number n must be part of a triple, first suppose that n is odd. Then n can be represented as the shorter leg as in the diagram below. Note that the longer leg is always one less than the hypotenuse, as in (3,4,5), (5,12,13), and (7,24,25).

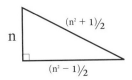

Now for the case where n is even. If n = 4, we already have a solution—namely, the (3,4,5) right triangle. If n is greater than or equal to 6, let n = m × k, with m odd. We can form a Pythagorean triple using m, as above. Then we simply multiply all the sides by k. (The final case, where n is a power of 2, is similar.)

2) 15 is the smallest such number. It is the hypotenuse of the 9–12–15 triangle (obtained by multiplying the 3–4–5 triangle by 3 on each side); it is the smaller leg of the 15–36–39 triangle (obtained by multiplying the 5–12–13 triangle by 3); and it is the larger leg of the 8–15–17 right triangle.

A Decent Decade (page 196)

The answer is the 1870s, because 1871, 1873, 1877, and 1879 are all prime numbers.

To prove this result, note first the obvious point that only those years ending in 1, 3, 7, or 9 are candidates for primality. Also note that each of these digits is congruent to either 0 or 1 (mod 3)—in other words, the remainder of the sum upon division by 3 must be 1. Therefore, the sum of the first three digits of the decade must be congruent to 1 (mod 3), because otherwise one of the four candidates in that decade would be divisible by 3.

From 1776 to the year 2000, there have been eight decades whose first three digits add up to be 1 (mod 3), as follows: 1780s, 1810s, 1840s, 1870s, 1900s, 1930s, 1960s, 1990s.

We can rule out all but one of these decades by simple trial and error, which produces the following results:

$$1781 = 13 \times 137$$
$$1813 = 7 \times 259$$
$$1841 = 7 \times 263$$
$$1903 = 11 \times 173$$
$$1939 = 7 \times 277$$
$$1969 = 11 \times 179$$
$$1991 = 11 \times 181$$

That leaves the 1870s as the only real candidate, and indeed it turns out that 1871, 1873, 1877, and 1879 are all prime, which admittedly can be deduced only laboriously or by computer. But if you take our word that there was such a decade, the above shows that the 1870s must be it.

Composite Sketch (page 196)

The first all-composite decade consisted of the years 200 through 209. Note that $201 = 3 \times 67$, $203 = 7 \times 29$, $207 = 9 \times 23$, and $209 = 11 \times 19$; the other years in the decade are composite since they are even or divisible by 5.

Five Squares to Two (page 197)

The diagram below shows where the cuts should be made. Note that the two pieces that must be moved to form the 1×2 rectangle are of the same size and shape.

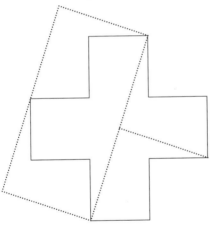

Equality, Fraternity (page 197)

1) $6^14^5 - 1 = 6143$

2) $-127 = 1 - 2^7$

Other solutions are possible.

Blue on Blue (page 198)

The smallest possible value of N is 10. Specifically, if there is one red ball and nine blue balls, the probability that the first selected ball is blue equals $9/10$; the probability that the second ball is blue equals $8/9$, and so on. The probability that all five balls are blue equals $(9/10)(8/9)(7/8)(6/7)(5/6) = 5/10 = 1/2$.

This same result can be proved without this sort of computation, but note that the product is trivial to calculate because all of the factors cancel out except for 5 and 10. In general, if there are n balls selected, the smallest value of N that produces a probability of $1/2$ for all n balls to be blue equals 2n.

Stick Figures (page 198)

A) $11 = -11 - 33 + 55$

B) $TT = 113 \overline{|355}$

Yes, that's π on the left hand side of the second equation. Remarkably, $355/113$ equals 3.14159292035..., whose first six decimal places match those of π (3.1415926535...).

Heralding Loyd (page 199)

The beauty of the puzzle is that you don't need to know anything about the boats' relative speeds to figure out the width of the river (although you can certainly deduce the relative speeds after obtaining the answer).

When the boats first meet, the total distance they have traveled equals the width of the river. By the time they meet again, the total distance traveled equals *three* times the width of the river. (Draw a diagram to convince yourself.) The boats are each traveling at a constant speed, so they each will have traveled three times as far by the second meeting as the distance they'd traveled by the first time they met. Because the boat starting in New York had traveled 720 yards at the first meeting, it must have traveled 2,160 yards at the time of the second meeting. But this distance is 400 yards from the other shore, so the width of the river equals 2,160 − 400 = 1,760 yards. Conveniently, this is exactly one mile.

Lots of Confusion (page 199)

If N is the number of lots and I is the original price per lot, we can obtain the following equations (each of the three expressions represents net profit):

$$18N − 243 = 6I = N(18 − I)$$

From the first and third equations we get $I = {}^{243}\!/_N$, and substituting into the second equation and combining with the first yields the equation $18N^2 − 243N − (6 \times 243) = 0$. Dividing by 9 produces $2N^2 − 27N − (6 \times 27) = 0$. This factors into $(2N + 9)(N − 18) = 0$, so we see that N = 18.

Oh, Henry! (page 200)

The answer is 21 square miles. The puzzle can be solved in an unusual manner by noticing some interesting facts about the areas given in the question: $388 = 8^2 + 18^2$ and $153 = 3^2 + 12^2$ and $61 = 5^2 + 6^2$. Using this information, you can construct the marvelous diagram below. Here, line segment A stands for one side of region A, etc. But the area of the

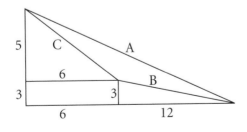

thin triangular region is simply the area of the big triangle—½ × 8 ×18—minus the areas of the two smaller triangles and rectangle (½ × 5 × 6, ½ × 3 × 12, and 3 × 6, respectively). This yields 72 − (15 + 18 + 18) = 72 − 51 = 21 square miles.

Walking the Blank (page 201)

The probability is one. No matter how you fill in the digits, the resulting 28-digit number will be divisible by 396.

To see why this works, note that 396 = 4 × 9 × 11. At this point we simply need to know the proper divisibility tests for 4, 9, and 11, which are stated below.

4: The last two digits form a number that is divisible by 4.

9: The sum of the digits must be divisible by 9.

11: The sum of the digits in odd positions minus the sum of the digits in even positions must be divisible by 11.

Note that the number ends in 76, which is divisible by 4, and therefore so is the entire number. And the sum of the existing 18 digits equals 90, which is divisible by 9, as is the sum of 0 through 9, so the whole number is divisible by 9. It only remains to check for divisibility by 11. The sum of the "odd" digits is 5 + 3 + 3 + 8 + 2 + 9 + 6 + 5 + 8 + 2 + 3 + 9 + 3 + 7 = 73, while the sum of the "even" digits is 8 + 3 + 0 + 6 + (sum of 1 through 9) = 8 + 3 + 0 + 6 + 45 = 62. The difference of 73 and 62 is 11, which is of course divisible by 11. The entire number must always be divisible by 396, no matter where the digits 0 through 9 are placed. This puzzle was originally created by Leo Moser—as an April Fool's prank!

Breaking the Hex (page 201)

The larger hexagon is precisely three times as big as the smaller hexagon. To see why, create the following diagram, where each of the six equilateral triangles in the original diagram is essentially flipped on its base, forming six more equilateral triangles in the middle. But each of the other six (isosceles) triangles has the same area as each of the equilateral triangles! (They have the same bases, and the altitude of any of the equilateral triangles, as dropped from one of the vertices of the large hexagon, is also an altitude for the isosceles triangle.) All in all, the large hexagon contains 18 triangles of the same area, and the small hexagon contains 6 such triangles, so the large hexagon is three times as big.

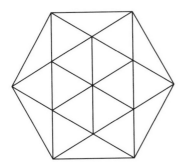

Domino Theory (page 202)

The probability is 17/48.

The key observation is that in order for a continuous chain of dominoes to be made in the manner described, the leftmost and rightmost spots must be the same. (If different, there would be no way for either to appear an even number of times, which is essential.) Now we make the teensy-weensy (but valid) assumption that for any pair of dominoes that share a spot, we can always construct a continuous chain with that spot at either end.

With that assumption under our belts, here is the complete set of ordered pairs that 1) contain at least one 5 and one 6, and 2) contain a pair of repeated spots to serve as the endpoints of the chain:

5–0 & 6–0	5–6 & 6–0	5–0 & 6–5
5–1 & 6–1	5–6 & 6–1	5–1 & 6–5
5–2 & 6–2	5–6 & 6–2	5–2 & 6–5
5–3 & 6–3	5–6 & 6–3	5–3 & 6–5
5–4 & 6–4	5–6 & 6–4	5–4 & 6–5
5–5 & 6–5		
5–6 & 6–6		

The total number of ordered pairs containing at least one 5 and at least one 6 equals 48 (7 × 7 minus the 5–6 domino, which can't be repeated), so the desired probability must be ¹⁷⁄₄₈.

The Wayward Three (page 202)

The number is 3,529,411,764,705,882, which when multiplied by ³⁄₂ produces 5,294,117,647,058,823.

To see how to get these numbers, we'll call them A and B, respectively. The first step in determining A is to see that its last digit must be 2. That's because ³⁄₂ ends in 3, so ⁄₂ must end in 1.

If A ends in 2, B must end in 23, because B is formed from A by putting the 3 at the end. But B = 3 × (⁄₂), so ⁄₂ must end in 41 (4 is the only digit whose product with 3 ends in 2). If ⁄₂ ends in 41, A ends in 82.

We keep going in this fashion, working right to left, until we (finally) get to a point where we encounter a 3, at which point the successive divisions end. That doesn't happen until the 16th digit!

Down to the Wire (page 203)

The probability of a four-game sweep by either team is $\frac{1}{2} \times \frac{1}{2} \times \frac{1}{2} \times \frac{1}{2} = \frac{1}{16}$, so overall the probability of a four-game Series is $\frac{1}{8}$ (because either team could win). The probability of a five-game Series—according to the information supplied in the puzzle—is $\frac{1}{4}$. Therefore the likelihood of the Series' going either six or seven games is $1 - (\frac{1}{8} + \frac{1}{4}) = \frac{5}{8}$. But the probability of a six-game Series must equal the probability of a seven-game Series, simply because once game 6 is reached, each team has a 50–50 shot at winning it! Therefore the probability of a seven-game Series is $\frac{5}{16}$.

Easy as A, B, C (page 203)

The numbers are 2, 3, and 6: $\frac{1}{2} + \frac{1}{3} + \frac{1}{6} = 1$. (This is the only solution with distinct integers. The other solutions are $\frac{1}{2} + \frac{1}{4} + \frac{1}{4} = 1$, and $\frac{1}{3} + \frac{1}{3} + \frac{1}{3} = 1$.)

The Beanpot Rally (page 203)

There are 48 possible combinations for the Beanpot tournament. To see why, note that any one school—say, BU—can have three possible opponents in the first round. After the match-ups are set, each game can end in one of two ways. As there are four games in all (including the third-place playoff), the total number of possibilities equals $3 \times 2 \times 2 \times 2 \times 2 = 48$. (Note that flipping the brackets, or flipping the schools within any one bracket, does not actually add to the possible outcomes.)

3, 4, 6, Hike! (page 204)

Let x = the total distance traveled, and y = the uphill (or downhill) distance. Because time = distance ÷ speed, the total time traveled is given by $2(((\frac{x}{2}) - y)/4) + \frac{y}{3} + \frac{y}{6}$, the factor of 2 reflecting the fact that the level portion of the trip was hiked twice.

Therefore $2(((\frac{x}{2}) - y)/4) + \frac{1}{3} + \frac{1}{6} = 5$, which looks like one equation with two unknowns (which would be unsolvable), but in fact the y terms all cancel out, leaving $\frac{x}{4} = 5$, or $x = 20$. The length of the trip was thus 20 miles.

Even Steven (page 205)

$79 + 5 + \frac{1}{3} = 84 + \frac{2}{6} + 0$

Top Score (page 205)

The product is maximized when the numbers are 3, 3, 3, 3, 3, 3, and 2. The maximum product is therefore 1,458.

A Bridge Too Far (page 205)

The answer is six points.

In order for a pair scoring 420 to have received 2½ points, the eight hands must have produced the following scores: 420, 420, 420, 420, 420, 420, 450, 450. (Remember, everyone playing the hand scored either 420 or 450, so the only way to achieve 2½ points would be to tie five other pairs.)

If the pair in question had scored 450 instead, there would be a total of five 420s and three 450s. A pair scoring 450 would get one point for each of the 420s and ½ a point for each of the other two 450s, for a total of six points.

Seven-point Landing (page 206)

The diagram on the next page does the trick, and it's not as arbitrary as it may appear. Construct a rhombus with one-inch sides having vertices A, B, C, and D. The distance from A to C should also be one inch. The rhombus is essentially made of two equilateral triangles glued together.

Now rotate the rhombus to the right (keeping it fixed at point B) so that point X is one inch away from point D. The seven vertices of the rhombi are the answer.

Check it out. It works!

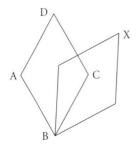

'Round Goes the Gossip (page 206)

The minimum number of phone calls required before everyone knows everything is 8.

Label the busybodies A, B, C, D, E, and F, and divide the six into two groups: A, B, and C are in Group 1; D, E, and F are in Group 2. One sequence of calls that does the job is as follows:

1) A–B
2) A–C (A and C now know all of Group 1)
3) D–E
4) D–F (D and F now know all of Group 2)
5) A–D (A and D now know everything)
6) C–F (C and F now know everything)
7) B–A (B knows everything)
8) E–F (E knows everything)

Note that in step 7, any one of A, C, D, and F could call B, and similarly for step 8. It turns out that 8 is the minimum number of phone calls necessary. In general, if you had n busybodies, the minimum number of calls required before everyone knows everything equals $2n - 4$.

Hitter's Duel (page 207)

Going 9 for 12 is equivalent to going 7 for 8 and then going 2 for 4. Assuming the two men had a comparable number of at bats before the games, their averages would still be close—perhaps identical—after both had gone 7 for 8. And as long as their averages were well below .500, Jackson's extra 2-for-4 increment could be enough to raise his average above Cobb's.

Visible and Divisible (page 207)

Note that $7 \times 8 \times 9 = 504$. Upon dividing 789,000 by 504, you get a remainder of 240. Therefore, if you add $504 - 240 = 264$ to 789,000, you get the desired number: 789,264. This answer is unique because the only other number of the form 789,XYZ to be divisible by 504 is 789,264 + 504 = 789,768, which repeats both the 7 and the 8.

Scale Drawing (page 207)

The equation linking the two scales is $F = (\frac{9}{5})C + 32$. To find the temperature that reads the same on both scales, just set F and C equal to one another, yielding $C = (\frac{9}{5})C + 32$. This gives $(\frac{4}{5})C = -32$, so $C = -40$. Therefore the only temperature that reads the same on both the Celsius and Fahrenheit scales is 40 degrees below zero.

Playing All the Angles (page 208)

The dissection can be completed as follows: The length of a side of triangle A must be ⅗ of a side of the original triangle, and a side of the three-piece triangle is ⅘ the original. The key is to cut the triangle in such a way that the shorter sides of piece B (bottom and right) are equal.

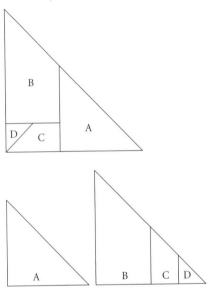

How Big? (page 208)

The easiest method is to draw a triangle within the wedge, as shown on the next page.

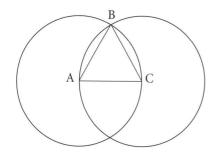

Note that triangle ABC must be equilateral, because each side is a radius of the circle, so angle BAC is 60 degrees. The pie-shaped figure determined by angle BAC (or angle BCA) must therefore have area $\pi/6$, because it represents one-sixth of a circle of radius one. The area of the upper half of the wedge we're interested in is the sum of those two sectors minus the area of the triangle, or $2(\pi/6) - \sqrt{3}/4$. The whole wedge thus has area $2\pi/3 - \sqrt{3}/2$.

Tunnel Division (page 209)

There are 5,280 feet in a mile and 3,600 seconds in an hour ($3,600 = 60 \times 60$). Therefore, 90 miles per hour is the same as $(90 \times 5280)/3600 = (9 \times 528)/36 = 528/4 = 132$ feet per second.

With this in mind, if the train takes four seconds to completely enter the tunnel, its length must be $132 \times 4 = 528$ feet. The length of the tunnel must be 132×40, or 5,280 feet. The tunnel is precisely one mile long!

Prime Time (page 209)

There are many different solutions. One, for example, is given by the sequence 0 7 4 3 2 5 6 1.

Triangle Equalities (page 209)

The three triangles are as follows:

Triangle	Area	Perimeter
12–16–20	96	48
10–24–26	120	60
9–40–41	180	90

Note that the first triangle is four times the 3–4–5 right triangle, while the second one is twice the 5–12–13 right triangle.

Dividing the Pentagon (page 209)

The solution is reached by forming another regular pentagon within the larger one, rotating the smaller one slightly, then joining the endpoints. The result is this figure:

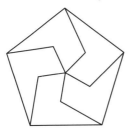

Planting the Sod (page 210)

Although 4444^{4444} is an enormous number, it gets substantially smaller upon taking the sum of its digits. To begin with, the number of digits in 4444^{4444} is less than the number of digits in 10000^{4444} (which equals $(10^4)^{4444} = 10^{17776}$), which has 17,777 digits. So SOD(4444^{4444}) < 17,777 $\times$ 9 = 159,993. Therefore SOD(SOD(4444^{4444})) < SOD(99,999), because the sum of the digits in 99,999 is at least as big as the SOD of any other number less than 159,993. Now SOD(99,999) = 45, so SOD(SOD(SOD(4444^{4444}))) is at most SOD(39)—again, 39 has the highest SOD of any number less than 45—and this equals 12.

But we also know that if we divide SOD(SOD(SOD(4444^{4444}))) by 9, the remainder is the same as the remainder upon dividing 4444^{4444} by 9, and, believe it or not, this can be calculated. We know that $4444 \equiv 7 \pmod 9$, and that $7^3 \equiv 343 \equiv 1 \pmod 9$, so $4444^{4444} \equiv 7^{4444} = 7 \times 7^{4443} = 7 \times (7^3)^{1481} \equiv 7 \pmod 9$. Therefore SOD(SOD(SOD($44444444$))) is a number that is less than 12 and is congruent to 7 (mod 9), so it must be 7.

Higher Than You Think (page 211)

The smallest number N such that it is impossible to create a dollar out of N coins is N = 77. That's because it is impossible to create 25 cents out of two coins; more generally, it is impossible to create 25 + 5x cents out of 2 + 5x coins.

Pocket Change (page 211)

Your friend must have 15 coins in his pocket. Here are the six ways that you can create a dollar from 15 coins. (Remember, no half-dollars allowed.)

	Q	D	N	P
1)	3	1	1	10
2)	2	1	7	5
3)	1	1	13	0
4)	1	5	4	5
5)	0	9	1	5
6)	0	5	10	0

Believe it or not, 15 is the only number such that there are precisely six ways of creating a dollar from that number of coins! The answer to the puzzle is therefore unique.

Square Not (page 212)

128 is the largest integer that cannot be expressed as the sum of distinct squares. Note that $128 - 100 = 28$, which is not expressible, and similarly for $128 - 81 = 47$, etc. For the record, here are some ways in which the integers from 129 through 150 can be expressed:

$129 = 100 + 25 + 4$
$130 = 100 + 25 + 4 + 1$
$131 = 81 + 49 + 1$
$132 = 81 + 25 + 16 + 9 + 1$
$133 = 81 + 36 + 16$
$134 = 81 + 36 + 16 + 1$
$135 = 100 + 25 + 9 + 1$
$136 = 100 + 36$
$137 = 100 + 36 + 1$
$138 = 100 + 25 + 9 + 4$
$139 = 100 + 25 + 9 + 4 + 1$

$140 = 100 + 36 + 4$
$141 = 100 + 36 + 4 + 1$
$142 = 100 + 25 + 16 + 1$
$143 = 81 + 36 + 25 + 1$
$144 = 144$
$145 = 144 + 1$
$146 = 100 + 36 + 9 + 1$
$147 = 81 + 36 + 25 + 4 + 1$
$148 = 144 + 4$
$149 = 100 + 49$
$150 = 100 + 49 + 1$

Dueling Weathermen (page 212)

If it is sunny, that means that WET was wrong and WILD was right, which occurs with probability $= (\frac{1}{4})(\frac{4}{5}) = \frac{1}{5}$.

If it rains, WET was right and WILD was wrong, which occurs with probability $= (\frac{3}{4})(\frac{1}{5}) = \frac{3}{20}$.

The odds of rain are therefore $\frac{3}{20}$ to $\frac{1}{5}$, or 3 to 4. Therefore the probability of rain equals $\frac{3}{(3+4)} = \frac{3}{7}$.

High Math at the 7–11 (page 213)

Note that $711 = 9 \times 79$, and 79 is prime. One of the items must therefore be a multiple of $0.79. Here are the first six multiples of $0.79, followed by $7.11 minus these numbers (the sum of the remaining three items) and $7.11 divided by these numbers (the product of the remaining three items).

	0.79	1.58	2.37	3.16	3.95	4.74
Sum	6.32	5.53	4.74	3.95	3.16	2.37
Prod.	9.00	4.50	3.00	2.25	1.80	1.50

We now must solve the problem of what three numbers add to the sum and multiply to the corresponding product. Only the pair (3.95, 2.25) looks at all friendly, because the others force us to create a product that is a round number from a sum that is not round. After a little trial and error, note that if you subtract (divide) the sum (product) number by 1.25, you get a required sum (for two numbers) of 2.70, and a corresponding product of 1.80. Now we're talking. We see that 1.50 and 1.20 add to 2.70 and multiply to 1.80. The four items therefore cost $3.16, $1.25, $1.50, and $1.20.

Who Am I? (page 213)

From statement 2, if I'm a multiple of 3, I must be one of 51, 54, or 57. But none of these is a multiple of 4, which contradicts statement 1. So I'm not a multiple of 3. If I'm not a multiple of 3, I can't be a multiple of 6, so by statement 3, I must be one of 71, 73, 74, 75, 76, 77, 79. But if I'm one of these numbers, then statement 1 says that I must be a multiple of 4. And the only multiple of 4 in that list is 76. So I am 76.

Home on the Range (page 214)

If 11 sheep can last 8 days, it would appear that the pasture has 88 "sheep-days" in it. But if 10 sheep can last 9 days, for a total of 90 sheep-days, we have to conclude that the extra day's growth amounts to two sheep-days. If so, two sheep could last their entire lives!

Heads or Tails (page 214)

The probability is precisely ½ that at some point there will be 3 or more consecutive flips that come out the same.

To see why, look first at the probability that three or more heads will come up. There are eight sequences with three consecutive heads, as follows: HHHTT, HHHTH, HHHHT, THHHT, THHHH, HTHHH, TTHHH, HHHHH.

Similarly, there are eight sequences with three or more tails. Putting them together, 16 of the 32 possible combinations involve three consecutive flips that are the same, so the probability of this event equals ½.

INDEX

Page key: puzzle, *solution*

3, 4, 6, Hike! **204,** *266*
Agent 86 **105,** *239*
Ah, Yes, I Remember It Well **162**
All About Pythagoras **195,** *258*
All Together Now **136**
An Additional Trick **150**
An Easy Square **167**
Antennas **64,** *227*
Apple Picking **108,** *241*
Architect Art **74,** *230*
As I Recall **159**
Average Student, The, **92,** *235*
Beanpot Rally, The, **203,** *266*
Big Difference **93,** *235*
Birthday Hugs **23,** *217*
Birthday Surprise, The, **110,** *242*
Blue on Blue **198,** *261*
Born Under a Bad Sign **185,** *247*
Breaking the Hex **201,** *264*

Bridge Too Far, A, **205,** *267*
Bundles of Tubes **67,** *228*
Cereal Serial **189,** *251*
Chewed Calculator **54,** *225*
Circular Logic **183,** *246*
Colored Balls #1 **80,** *232*
Colored Balls #2 **81,** *232*
Comic Relief **105,** *240*
Composite Sketch **196,** *260*
Cookie Jars **46,** *224*
Crackers! **35,** *220*
Crate Expectations **36,** *221*
Crushed Calculator **55,** *226*
Cube of Cheese **32,** *219*
Cubes & Squares **57,** *226*
Cutting the Horseshoe **71,** *230*
Cyclomania **42,** *222*
Decent Decade, A, **196,** *259*
Diamond in the Rough **100,** *237*
Disappearing Apples **79,** *232*
Divide and Conquer **104,** *239*

Dividing the Pentagon **209,** *272*
Domino Theory **202,** *264*
Don't Make My Brown Eyes Blue
    186, *248*
Donut Try This at Home **95,** *236*
Double Trouble **189,** *251*
Down to the Wire **203,** *266*
Dueling Weathermen **212,** *274*
Easier by the Dozen **90,** *234*
Easy as A, B, C **203,** *266*
Easy Way, The, **125**
Equality, Fraternity **197,** *260*
Even Steven **205,** *267*
Find the Gold **51,** *225*
First Class Letters **188,** *250*
Five Squares to Two **197,** *260*
Fleabags **47,** *224*
Floating Family **26,** *218*
French Connection, The **112,** *242*
Frisky Frogs **52,** *225*
From Start to Finish **107,** *240*
Generation Gap **109,** *242*
Getting Along **131**
Gloves Galore **22,** *217*
Going Crackers **114,** *243*
Going Off on a Tangent **194,** *256*
Good Neighbor Policy **70,** *230*
Heads or Tails **214,** *276*
Heralding Loyd **199,** *261*
High Math at the 7-11 **213,** *275*
Higher Than You Think **211,** *273*

High-Speed Copying **103,** *239*
Hitter's Duel **206,** *269*
Home on the Range **214,** *276*
House Colors **62,** *227*
How Big? **208,** *270*
Hummer by Phone **128**
It All Adds Down **135**
Leaping Lizards **53,** *225*
LoadsaLegs **63,** *227*
Logical Pop **82,** *232*
Long and the Short of the Grass,
    The, **28,** *218*
Long Division **98,** *237*
Long String, The, **187,** *249*
Long Way Around, The, **96,** *237*
Lots of Confusion **199,** *262*
Magic Hexagon **89,** *234*
Magic Triangle **88,** *233*
Mathbits
    18 and 81 **33**
    Add Up to Squares **18**
    Calculator Trick **45**
    Circumference of a Glass **26**
    Combination Locks **50, 59**
    Cross Out Divide **41**
    Divide 100 by 81 **68**
    Divisible by 3? **73**
    Equals Sign **46**
    Fifty/Fifty Chance **62,** *217*
    Five Odd Figures **70,** *220*
    Four 9s **42,** *232*

Galileo's Pendulums **79**
Largest 2-digit Number **31**
Perfect Numbers **64**
Pi Reminder **29**
Prime Number? **82**
Rule of Thumb **23**
Sevenths Patterns **48**
Spider Web **53**
X-axis Reminder **28**
Mirror Time **113,** *243*
Missing Shekel, The, **115,** *243*
Missing Six, The, **94,** *236*
Mnemonics **11–15**
Most Valuable Puzzle **188,** *250*
Multisox **72,** *230*
Nickel For Your Thoughts, A, **138**
Nine Coins **29,** *218*
No Burglars! **75,** *231*
No Calculators, Please **185,** *246*
Not Such a Big Difference **93,** *235*
Odd Balls **31,** *219*
Oh, Henry! **200,** *262*
Old MacDonald **58,** *226*
Old Mrs. MacDonald **59,** *226*
On the Square **171**
On the Trail **107,** *240*
One, Two, Three **186,** *248*
Page Boy **184,** *245*
Pairing Off **190,** *251*
Pencil Squares **40,** *222*
Pencil Triangles **41,** *222*

Perforation! **78,** *231*
Performance Anxiety **189,** *251*
Picnic Mystery **50,** *225*
Pieces of Eight **106,** *240*
Pizza and the Sword **39,** *221*
Planting the Sod **210,** *272*
Playing All the Angles **208,** *270*
Playing the Triangle **109,** *241*
Pocket Change **211,** *273*
Potato Pairs **33,** *219*
Power of Seven Continues, The,
    66, *228*
Power of Seven, The, **65,** *228*
Powers of Four, The, **102,** *239*
Prime Time **209,** *271*
Puzzle of the Sphinx **77,** *231*
Pyramids **68,** *229*
Quarter Horses **116,** *244*
Rolling Quarter, The, **48,** *224*
Roman Knows, A, **184,** *245*
'Round Goes the Gossip **206,** *268*
Run-off, The, **111,** *242*
Scale Drawing **207,** *269*
Secret Number Codes **85,** *233*
Sesquipedalian Farm **44,** *223*
Seven-point Landing **206,** *267*
Shape Code **86,** *233*
Shutting the Eye **192,** *253*
Sliding Quarters **49,** *224*
Slippery Slopes **27,** *218*
Sly Inference, A, **121**

Sneaky Serpent, The, **138**

So Where's the Money? **176**

Sox Unseen **21,** *217*

Speedy Adder, The, **143**

Spring Flowers **43,** *223*

Square Deal, A, **190,** *252*

Square Not **212,** *274*

Square Route **94,** *236*

Squares & Cubes & Squares **70,** *229*

Squares & Cubes **56,** *226*

Stamp Collection, The, **187,** *248*

Staying in Shape **113,** *243*

Stick Figures **198,** *261*

Sticky Shakes **24,** *218*

Sugar Cubes **34,** *220*

Surprise Ending **195,** *257*

Survival of the Splittest **191,** *253*

Tennis Tournament **61,** *227*

Thanksgiving Feast **195,** *257*

Three Js **73,** *230*

Three's a Charm **101,** *238*

Three-Quarters Ranch **45,** *223*

Thumbs Down **193,** *255*

Ticket to Ride **194,** *255*

Too Close to Call **187,** *249*

Top Score **205,** *267*

Train Crash **76,** *231*

Triangle Equalities **209,** *272*

tricks for multiplying **16–18**

tricks for remembering **11–18**

Tricky Connections **30,** *219*

Tunnel Division **209,** *271*

Very Good Year, A, **97,** *237*

Visible and Divisible **207,** *269*

Walking the Blank **201,** *263*

Wayward Three, The, **202,** *265*

What's in a Name? **192,** *254*

When in Rome **99,** *237*

Who Am I? **213,** *275*

Who Is Faster? **91,** *234*

Who Is the Liar? **102,** *238*

Wiener Triangles **60,** *227*

Witches' Brew **37,** *221*

Witches' Stew **38,** *221*

Wolf, the Goat and the Cabbage, The, **25,** *218*

Wrong Envelope? **69,** *229*